D0088098

A Photographic Atlas

for the

Botany Laboratory

FIFTH EDITION

Samuel R. Rushforth
Utah Valley University

Robert R. Robbins
Utah Valley University

John L. Crawley

Kent M. Van De Graaff
Weber State University

SUFFOLK UNIVERSITY
MILDRED F. SAWYER LIBRARY
8 ASHBURTON PLACE
BOSTON. MA 02108

Morton Publishing Company
925 W. Kenyon, Unit 12
Englewood, Colorado 80110

To biologists, conservationists,
and concerned people throughout the world
who actively strive to save nature.

Haleakala Silversword (*Argyroxiphium sandwicense*)
The Haleakala Silversword is only found in Haleakala National Park
on the Hawaiian island of Maui at elevations above 2100 meters.

Copyright 1994, 1995, 1998, 2004, 2008 by Morton Publishing Company

ISBN10: 0-89582-770-0
ISBN13: 978-0-89582-770-8

10 9 8 7 6 5 4 3 2

All rights reserved. Permission in writing must be obtained from the publisher before any part of this work may be reproduced or transmitted in any form, or by any means, electronic or mechanical, including photocopying and recording or by any information storage or retrieval system.

Printed in the United States of America

Preface

Botany, a specialty in biology, is the study of plants. More than 95% of the Earth's biomass is composed of plants. Because plants are so abundant, visible, and necessary to life, everyone has some interest in and knowledge of botany.

Plants and flowers, for example, are welcomed into our homes for their beauty. Florists remind us to "Say it with flowers." Flowers are easily cared for in a vase of water, where they remain attractive and aromatic for a few days. In contrast, plants are not so easily cared for, but they can live and grow for more than just a few days. House plants need to be properly potted, watered regularly, and occasionally fertilized. By providing this care, we learn about the nature and requirements of plants as we become amateur botanists. Grasses, trees, and shrubs landscaped around our houses provide us with beauty, comfort, shade, and a sense of being connected to nature.

Our existence depends on plants. From plants we are supplied with building materials, the oxygen we breathe, and the food we eat. Through photosynthesis, plants use sunlight to convert water and carbon dioxide to sugars, releasing oxygen as a by-product.

Plants also provide us with materials required for producing medicines and paper. Insights and discoveries in botany are occurring today at a rapid pace. What students will learn in a basic botany course will be of immeasurable value in understanding and making decisions about the ecological problems currently facing our world. A basic botany course is essential as a secure foundation for planning advanced life science study, as well as for basic scientific literacy.

Botany is a visually oriented science. *A Photographic Atlas for the Botany Laboratory* provides clear photographs and drawings of tissues and organisms, similar to specimens seen in most botany laboratories. It is designed to accompany any botany (or biology) text or laboratory manual used in the classroom. In certain courses this atlas could serve as the laboratory manual.

This atlas provides a visual representation of the major groups of botanical organisms. Care has been taken to construct labeled, informative figures. Parts of organisms are depicted clearly and accurately. The terminology used matches college botany texts.

Several dissections of plants are provided for students who have the opportunity to do similar dissections in the class laboratory. In addition, photomicrographs, photos of living specimens, and herbarium collections are included. These figures enhance the student's understanding of plant structure and plant classification. Plants of significant economic importance for food, shelter, and medicines are highlighted.

Preface to Fifth Edition

The success of the previous editions of *A Photographic Atlas for the Botany Laboratory* provided opportunities to make extensive changes to enhance the value of this new edition. The extensive revision of this atlas presented in its fifth edition required an inordinate amount of planning, organization, and work. As authors, we have the opportunity and obligation to listen to the critiques and suggestions from students and faculty who have used this atlas. This constructive input is appreciated and has resulted in a greatly improved atlas.

One objective in preparing this edition of the atlas was to create an inviting pedagogy. The page layout was improved by careful selection of new and replacement photographs. Seven new life cycles were added. Each image in this atlas was carefully evaluated for its quality, effectiveness, and accuracy. In all over 200 new images were included. Perhaps most important to this fifth edition was Robert Robbins, Utah Valley University. Bob has brought important professional input, and rounded out the team. We are indebted to Douglas Morton and the personnel at Morton Publishing Company for this opportunity and their encouragement and support.

Acknowledgments

Many professionals helped us prepare this atlas and share our enthusiasm for its value for botany students. We especially appreciate Brigham Young University and The University of Utah for allowing us to photograph specimens in their herbarium and botanical greenhouses. Kaye H. Thorne and Thomas G. Black were helpful in selecting specimens to include. Wilford M. Hess and James V. Allen provided scanning electron micrographs, and we appreciate their generosity. Dawn Gatherum, Eugene Bozniak, and Rachael Bush in the Department of Botany at Weber State University were generous in providing photographs and access to the university's botanical greenhouse. The suggestions and assistance of H. Blaine Furniss were greatly appreciated.

The professional input by reviewers and users has been invaluable. Margaret Olive, Pensacola Junior College, Brenda L. Young, Daemen College and Larry St. Clair, Brigham Young University, were most generous in their meticulous critique of the manuscript. Others who offered especially helpful input include Neil A. Harriman, Theodore Esslinger, Frank W. Ewers, Patrick F. Fields, Dale M. J. Mueller, John W. Taylor, Brian Speer, Lawrence Virkaitis, Cecile Bochmer, and Anne S. Viscomi.

Book Team

Publisher: Doug Morton
Managing Editor: David Ferguson
Typography and Design: Focus Design
Cover: Bob Schram, Bookends Publication Design
Illustrations: Jessica Ridd

Egyptian blue lotus (*Nymphaea caerulea*)
flowers open in the morning and close by afternoon. This
flower was highly significant in Egyptian mythology.

Table of Contents

Prelude

Scientists work to determine accuracy in understanding the relationship of organisms even when it requires changing established concepts. Development, structure, function, DNA sequence, the fossil record and geological dating are used to establish systematics and classify organisms. As new techniques become available, they too aid in our understanding of evolutionary relationships between groups of organisms and closely related species.

In 1758 Carolus Linnaeus, a Swedish naturalist, assigned all known kinds of organisms to two kingdoms–plants and animals. For over two centuries, this dichotomy of plants and animals served biologists well. In 1969, Robert H. Whittaker convincingly made a case for a five-kingdom system comprised of Monera, Protista, Fungi, Plantae, and Animalia. The five-kingdom basis of systematics prevailed for over twenty years, and is now being challenged with a new system that includes three domains (superkingdoms) and four kingdoms (see exhibit 1). This system, which is used in this edition of *A Photographic Atlas for the Botany Laboratory*, is based on criteria used in the past and new techniques in molecular biology. It is important to note, however, that a classification scheme is a human construct subject to alteration as additional knowledge is obtained.

Exhibit I Domains, Kingdoms, and Representative Examples

Bacteria Domain– Cyanobacteria, gram-negative and gram-positive bacteria

Archaea Domain– Methanogens, halophiles, and thermophiles

Eukarya Domain– Eukarya, single-celled and multicelled organisms; fungi, protists, plants, and animals

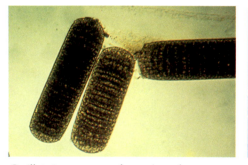

Oscillatoria sp., a cyanobacterium that reproduces through fragmentation

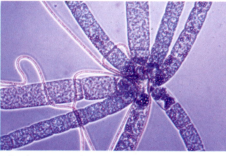

Thiothrix sp., a thermophile that oxidizes H_2S for an energy source

| Kingdom Fungi | Kingdom Protista | Kingdom Plantae | Kingdom Animalia |

Aspergillus sp., a mold that reproduces asexually and sometimes sexually

Volvox sp., a motile green alga that reproduces asexually or sexually

Musa sp., the banana, is high in nutritional value and is one of the twelve most important human food plants

Chamaeleo calyptratus, the veiled chameleon, is known for its ability to change colors according to its surroundings

Basic Characteristics of Domains

Domain	Characteristics
Bacteria Domain—Bacteria	Prokaryotic cell; single circular chromosome; cell wall containing peptidoglycan; chemosynthetic autotrophs, chlorophyll-based photosynthesis, photosynthetic autotrophs, and heterotrophs; gram-negative and gram-positive forms; lacking nuclear envelope; lacking organelles and cytoskeleton
Archaea Domain—Archaea	Prokaryotic cell; single circular chromosome; cell wall with peptidoglycan absent; membrane lipids, unique ribosomal RNA sequences; lacking nuclear envelope; some with chlorophyll-based photosynthesis; with organelles, and cytoskeleton
Eukarya Domain—Eukarya	Single-celled and multicellular organisms; nuclear envelope enclosing more than one linear chromosome; membrane-bound organelles in most; some with chlorophyll-based photosynthesis; cytoskeleton present

Common Classification System of Selective Living Eukaryotes

Eukarya Domain– eukaryotes
    Kingdom Fungi– fungi
        Phylum Zygomycota– bread molds and fly fungi
        Phylum Ascomycota– yeasts, molds, morels and truffles
        Phylum Basidiomycota– mushrooms, rusts and smuts
    Kingdom Protista– heterotrophic and photosynthetic protists
        Phylum Myxomycota– plasmodial slime molds
        Phylum Dictyosteliomycota (=Acrasiomycota)– cellular slime molds
        Phylum Oomycota– water molds
        Phylum Euglenophyta– euglenoids
        Phylum Cryptophyta– cryptomonads
        Phylum Rhodophyta– red algae
        Phylum Dinophyta (=Pyrrhophyta)– dinoflagellates
        Phylum Haptophyta– haptophytes
        Phylum Chrysophyta– golden algae
        Phylum Bacillariophyta– diatoms (diatoms are often placed in the phylum Chrysophyta)
        Phylum Phaeophyta– brown algae
        Phylum Chlorophyta– green algae
    Kingdom Plantae– bryophytes and vascular plants
        Phylum Hepatophyta– liverworts
        Phylum Anthocerophyta– hornworts
        Phylum Bryophyta– mosses
        Phylum Psilotophyta (=Psilophyta)– wisk ferns
        Phylum Lycophyta (=Lycopodiophyta)– club moss, ground pines and spike mosses
        Phylum Equisetophyta (=Sphenophyta)– horsetails
        Phylum Pterophyta (=Polypodiophyta)– ferns
        Phylum Cycadophyta– cycads
        Phylum Ginkgophyta– Ginkgo
        Phylum Pinophyta (= Coniferophyta)– conifers
        Phylum Gnetophyta– gnetophytes
        Phylum Magnoliophyta (=Anthophyta)– angiosperms (flowering plants)
    Kingdom Animalia– invertebrate and vertebrate animals (not discussed in this atlas)

All organisms are comprised of one or more cells. Cells are the basic structural and functional units of organisms. A cell is a minute, membrane-enclosed, protoplasmic mass consisting of chromosomes surrounded by cytoplasm. Specific organelles are contained in the cytoplasm that function independently but in coordination with one another. Prokaryotic cells and eukaryotic cells are the two basic types.

Eukaryotic cells contain a true *nucleus* with multiple chromosomes, have several types of specialized *organelles,* and have a differentially permeable cell (plasma) membrane. Organisms comprised of eukaryotic cells include protozoa, fungi, algae, plants, and invertebrate and vertebrate animals.

The more primitive *prokaryotic cells* lack a membrane-bound nucleus, instead they contain a single molecule of *DNA.* These cells have few organelles. A rigid or semi-rigid *cell wall* provides shape to the cell outside the *cell (plasma) membrane.* Bacteria are examples of prokaryotic, single-celled organisms.

The *nucleus* is the large, spheroid body within the eukaryotic cell that contains the genetic material of the cell. The nucleus is enclosed by a double membrane called the *nuclear membrane,* or *nuclear envelope.* The *nucleolus* is a dense, nonmembranous body in the nucleus composed of protein and RNA molecules. The chromatin is comprised of fibers of protein and DNA molecules. Prior to cellular division, the chromatin shortens and coils into rod-shaped *chromosomes.* Chromosomes consist of DNA and structural proteins called *histones.*

The *cytoplasm* of the eukaryotic cell is the medium between the nuclear membrane and the cell membrane. *Organelles* are small membrane-bound structures within the cytoplasm (other than the nucleus). The structure and functions of the nucleus and principal plant organelles are listed in Table 1.1. In order for cells to remain alive, metabolize, and maintain *homeostasis,* cells must have access to nutrients and respiratory gases, be able to eliminate wastes, and be in a constant, protective environment.

Plant cells differ in some ways from other eukaryotic cells in that their cell walls contain *cellulose* for stiffness. Plant cells also have vacuoles for water storage and membrane-bound *chloroplasts* with photosynthetic pigments for photosynthesis.

Tissues are groups of similar cells that perform specific functions. A flowering plant, for example, is composed of three tissue systems:

1. The *ground tissue system*, providing support, regeneration, respiration, photosynthesis, and storage;
2. The *vascular tissue system*, providing conduction of water, nutrients, and sugars through the plant;
3. The *dermal tissue system*, providing surface covering and protection.

Organs are two or more tissue systems that carry out specific functions together. Examples of organs include floral parts, leaves, stems, and roots.

The *organism* is the plant itself, which consists of all the organs functioning together to keep it alive, allow it to grow, and permit it to propagate.

The term *cell cycle* refers to how an organism develops, grows, and maintains and repairs body tissues. In the cell cycle, each new cell receives a complete copy of all genetic information in the parent cell, and the cytoplasmic substances and organelles to carry out hereditary instructions.

The cell cycle of a plant consists of growth, synthesis, mitosis, and cytokinesis. *Growth* is the increase in cellular mass as the result of metabolism. *Synthesis* is the production of DNA and RNA to regulate cellular activity. *Mitosis* is the exact duplication and division of chromosomes. *Cytokinesis* is the division of the cytoplasm that follows mitosis.

Unlike animal cells, plant cells have a rigid cell wall that does not cleave during cytokinesis. Instead, a new cell wall develops between the daughter cells. Furthermore, many land plants do not have centrioles. The *microtubules* in these plants form a barrel-shaped anastral *spindle* at each pole. Mitosis and cytokinesis in plants occurs in basically the same sequence as these processes in animal cells.

Asexual reproduction is propagation of new organisms without sex; that is, the production of new individuals by processes that do not involve *gametes.* Asexual reproduction occurs in a variety of microorganisms, plants, and animals, wherein a single parent produces offspring genetically identical to itself. Asexual reproduction is not dependent on the presence of other individuals. Neither meiosis nor gametes is required. In asexual reproduction, all the offspring are genetically identical (except for mutants). Types of asexual reproduction and example organisms include:

1. *fission*—a single cell divides to form two separate cells (bacteria, protozoans, and other one-celled organisms);
2. *sporulation*—many cells are formed that may remain separate or join together in a cyst-like structure (algae, fungi, protozoans);
3. *budding*—buds develop on the parent and then detach themselves (hydras, yeast, certain plants);
4. *fragmentation*—organisms break into two or more parts, and each part is capable of becoming a complete organism (flatworms, echinoderms, algae, some plants, and others).

Sexual reproduction is production of new organisms through the union of genetic material from two parents. Sexual reproduction usually involves meiotic nuclear division of diploid cells to produce haploid gametes (such as sperm and egg cells). This is followed by the fusion of the gametes during fertilization and the formation of a zygote.

The major biological difference between sexual and asexual reproduction is that sexual reproduction produces genetic variation in the offspring. The combining of genetic material from the gametes produces offspring that are genetically different from either parent and contain new combinations of genes. This may increase the ability of the species to survive environmental changes or to reproduce in new habitats. The only genetic variation that can arise in asexual reproduction comes from mutations.

Table 1.1 Structure and Function of Components of Eukaryotic Plant Cells

Component	Structure	Function
Cell (plasma) membrane	Composed of protein and phospholipid molecules	Provides form to cell; controls passage of materials into and out of cell
Cell wall	Cellulose fibrils	Provides structure and rigidity to plant cell
Cytoplasm	Fluid to jelly-like substance	Serves as suspending medium for organelles and dissolved molecules
Endoplasmic reticulum	Interconnecting membrane-lined channels	Enables cell transport and processing of metabolic chemicals
Ribosome	Granules of nucleic acid (RNA) and protein	Synthesizes protein (translation)
Mitochondrion	Double-membraned sac with cristae (inner membrane folds)	Assembles ATP (cellular respiration)
Golgi complex	Flattened membrane-lined chambers	Synthesize carbohydrates and packages molecules for secretion
Lysosome	Membrane-surrounded sac of enzymes	Digests foreign molecules and worn cells
Centrosome	Mass of protein that may contain rod-like centrioles	Organizes spindle fibers and assists mitosis and meiosis, centrioles form flagellar basal bodies
Vacuole	Membranous sac	Stores and releases substances within the cytoplasm, regulates cellular turgor pressure
Microfibril and microtubule	Protein strands and tubes	Forms cytoskeleton, supports cytoplasm and transports materials
Cilium and flagellum	Cytoplasmic extensions from cell; containing microtubules	Movement of particles along cell surface or cell movement
Nucleus	Nuclear envelope (membrane), nucleolus, and chromatin (DNA)	Contains genetic code that directs cell activity; forms ribosomes
Chloroplast	Double-membraned sac containing grana (thylakoid membranes)	Carries out photosynthesis

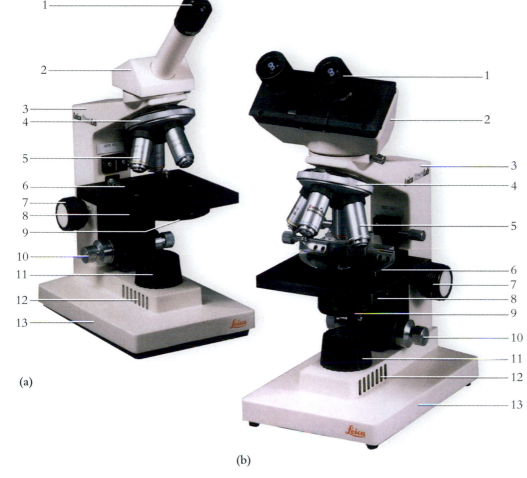

Figure 1.1 (a) A compound monocular microscope, and (b) a compound binocular microscope.

1. Eyepiece (ocular)
2. Body
3. Arm
4. Nosepiece
5. Objective
6. Stage clip
7. Focus adjustment knob
8. Fixed stage
9. Condenser
10. Fine focus adjustment knob
11. Collector lens with field diaphragm
12. Illuminator (inside)
13. Base

(a)

(b)

Photograph courtesy of: Leica Inc., Deerfield, IL

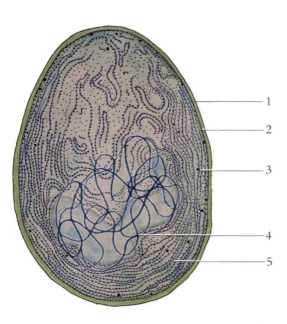

Figure 1.2 A prokaryotic cell.

1. Cell wall
2. Cell (plasma) membrane
3. Ribosomes
4. Circular molecule of DNA
5. Thylakoid membranes

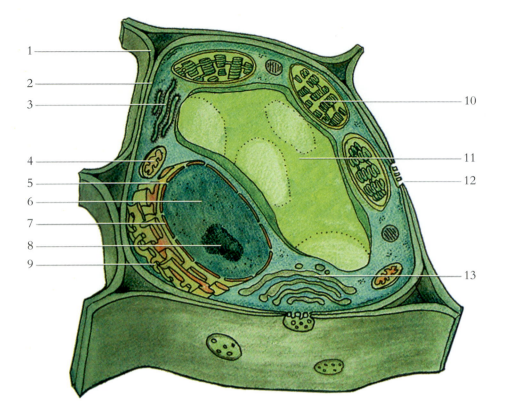

Figure 1.3 Diagram of a typical plant cell.
1. Cell wall
2. Cell (plasma) membrane
3. Rough endoplasmic reticulum
4. Mitochondrion
5. Nuclear pore
6. Nucleus
7. Nuclear membrane (envelope)
8. Nucleolus
9. Smooth endoplasmic reticulum
10. Chloroplast
11. Vacuole
12. Plasmodesmata
13. Golgi complex

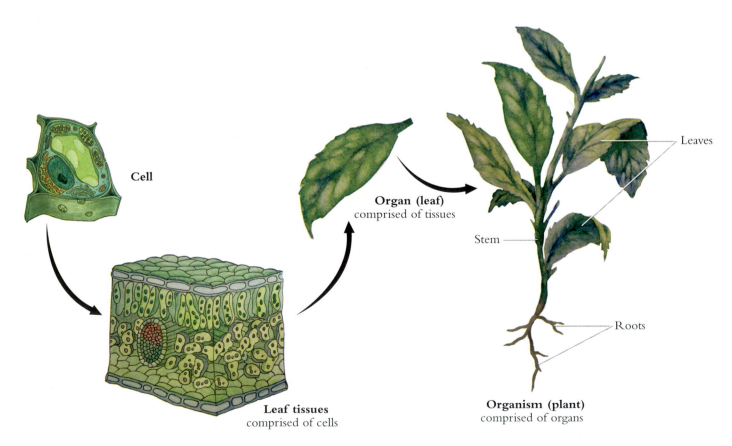

Cell

Leaf tissues
comprised of cells

Organ (leaf)
comprised of tissues

Leaves

Stem

Roots

Organism (plant)
comprised of organs

Figure 1.4 Structural levels of plant organization.

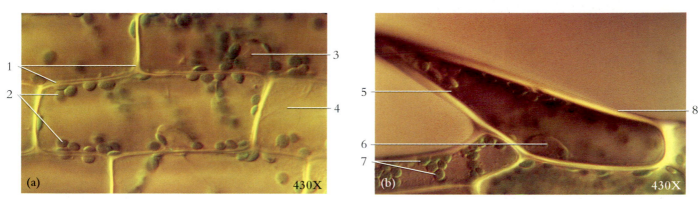

Figure 1.5 Live *Elodea* sp. leaf cells (a) photographed at the center of the leaf and (b) at the edge of the leaf.

1. Cell wall	3. Nucleus	5. Spine-shaped cell on	6. Nucleus	8. Cell wall
2. Chloroplasts	4. Vacuole	exposed edge of leaf	7. Chloroplasts	

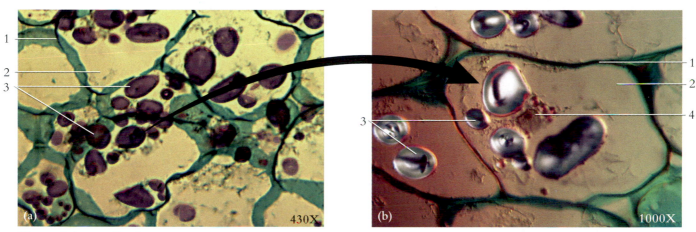

Figure 1.6 (a) Cells of a potato, *Solanum tuberosum*, showing starch grains at a low magnification, and (b) at a high magnification. Food is stored as starch in potato cells, which is deposited in organelles called amyloplasts.

1. Cell wall   2. Cytoplasm   3. Starch grains   4. Nucleus

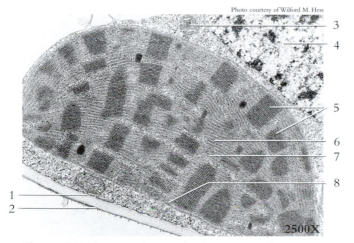

Figure 1.7 Electron micrograph of a portion of a sugar cane leaf cell.

1. Cell membrane   5. Grana
2. Cell wall   6. Stroma
3. Mitochondrion   7. Thylakoid membrane
4. Nucleus   8. Chloroplast envelope

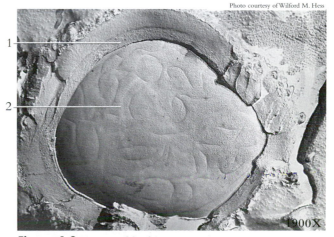

Photo courtesy of Wilford M. Hess

Figure 1.8 Fractured barley smut spore.

1. Cell wall   2. Cell membrane

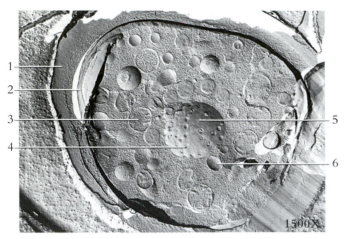

Figure 1.9 Barley smut spore, fractured through the middle of the cell.

1. Cell wall   4. Nuclear pore
2. Cell membrane   5. Nucleus
3. Mitochondrion   6. Vacuole

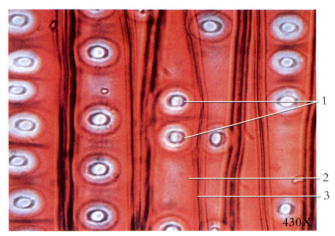

Figure 1.10 Longitudinal section through the xylem of pine, *Pinus* sp., showing tracheid cells with prominent bordered pits.

1. Bordered pits   3. Cell wall
2. Tracheid cell

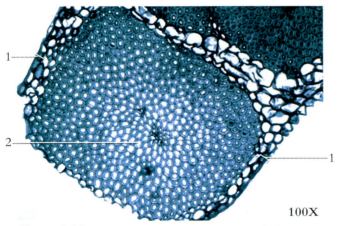

Figure 1.11 Transverse section through the leaf of a yucca, *Yucca brevifolia.* A bundle of leaf fibers (sclerenchyma) is evident at the edge of the leaf.

1. Epidermis   2. Sclerenchyma tissue

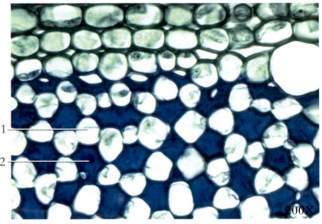

Figure 1.12 Collenchyma tissue from the stem of a begonia. Collenchyma tissue consists of thickened primary cell walls that form supportive strands beneath the epidermis in stems and petioles.

1. Cell lumen   2. Thickened primary cell wall

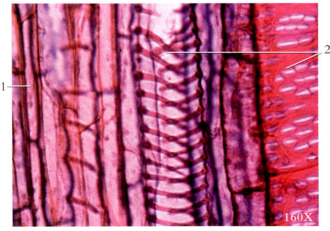

Figure 1.13 Longitudinal section through the xylem of a squash stem, *Cucurbita maxima.* The vessel elements shown here have several different patterns of wall thickenings.

1. Parenchyma   2. Vessel elements

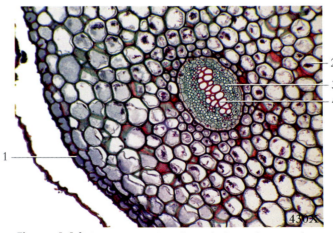

Figure 1.14 Transverse section through the underground stem (rhizome) of the fern *Polypodium* sp., showing multiple cell types.

1. Epidermis   3. Phloem sieve cells
2. Parenchyma   4. Xylem tracheid

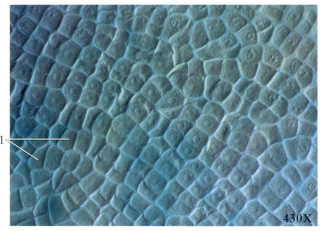

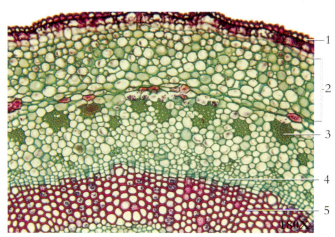

Figure 1.15 Surface view of the green alga, *Enteromorpha* sp., with individual cells pressed together to form a parenchyma-like tissue.

1. Cells of *Enteromorpha*

Figure 1.16 Transverse section of a stem from *Hoya carnosa*, wax plant.

1. Epidermis
2. Parenchyma (cortex)
3. Sclereids
4. Secondary phloem
5. Secondary xylem

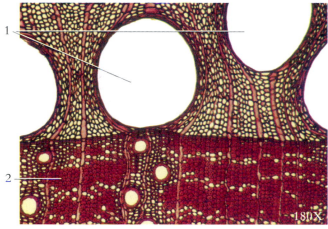

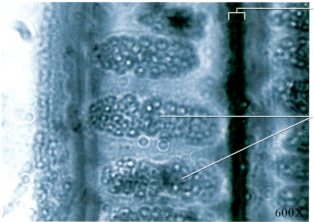

Figure 1.17 Transverse section of an oak, *Quercus* sp., stem through the secondary xylem (wood).

1. Vessel elements
2. Sclerenchyma (fibers)

Figure 1.18 Close-up of sieve tube elements in the phloem of a grape, *Vitis vinifera*. Note the sieve areas on the sieve tube elements.

1. Cell wall
2. Sieve areas

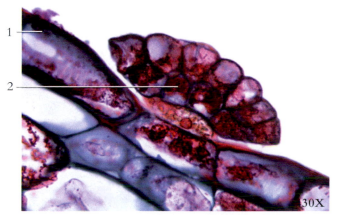

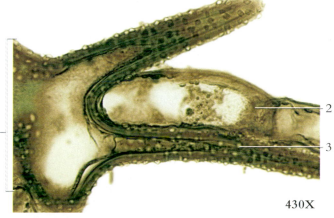

Figure 1.19 Section through a leaf of the venus flytrap, *Dionaea muscipula*, showing epidermal cells with a digestive gland. The gland is comprised of secretory parenchyma cells.

1. Epidermis
2. Gland

Figure 1.20 Astrosclereid in the petiole of a pondlily, *Nuphar* sp.

1. Astrosclereid
2. Parenchyma cell
3. Crystals in cell wall

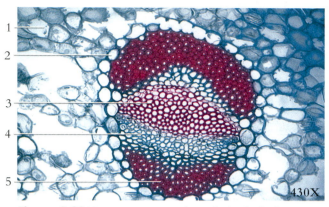

430X

Figure 1.21 Transverse section through the leaf of a yucca, *Yucca brevifolia*, showing a vascular bundle (vein). Note the prominent sclerenchyma tissue forming caps on both sides of the bundle.

1. Leaf parenchyma   4. Phloem
2. Bundle sheath   5. Bundle cap
3. Xylem

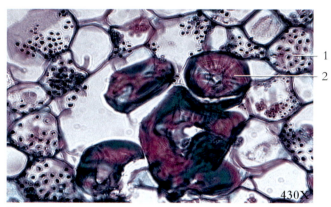

1100X

Figure 1.22 Section through the endosperm tissue of a persimmon, *Diospyros virginiana*. These thick-walled cells are actually parenchyma cells. Cytoplasmic connections, or plasmodesmata, are evident between cells.

1. Plasmodesmata   2. Cell lumen (interior space)

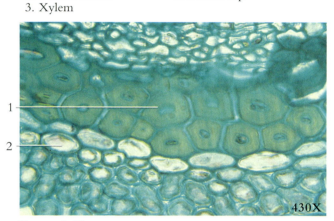

430X

Figure 1.23 Transverse section through the stem of flax, *Linum* sp. Note the thick-walled fibers as compared to the thin-walled parenchyma cells.

1. Fibers   2. Parenchyma cell

430X

Figure 1.24 Section through the stem of a wax plant, *Hoya carnosa*. Thick-walled sclereids (stone cells) are evident.

1. Parenchyma cell   2. Sclereid (stone cell)
   containing starch grains

Photo courtesy of B. A. Tait

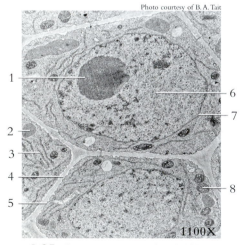

1100X

Figure 1.25 Electron micrograph of alfalfa root cells.

1. Nucleolus   5. Cell membrane
2. Immature plastid   6. Nucleus
3. Endoplasmic reticulum 7. Nuclear envelope
4. Cell wall   8. Mitochondrion

One (duplicated) chromosome composed of two identical chromatids

Chromatin strand

Centromere

Chromatid

Figure 1.26 Each duplicated chromosome consists of two identical chromatids attached at the centrally located and constricted centromere.

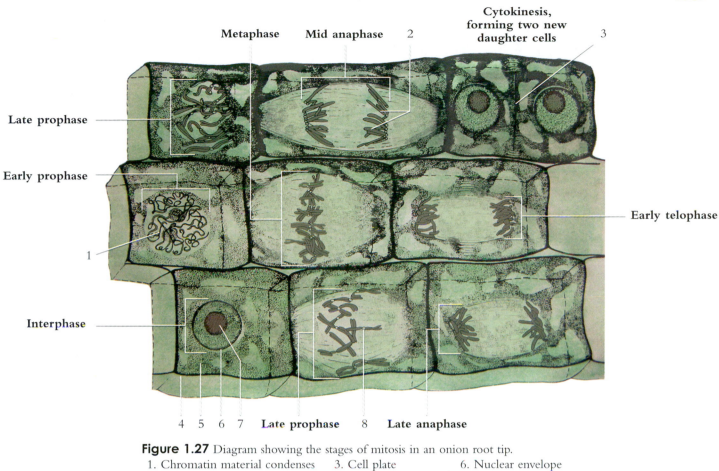

Figure 1.27 Diagram showing the stages of mitosis in an onion root tip.

1. Chromatin material condenses to form chromatids
2. Chromatids
3. Cell plate
4. Cell wall
5. Cytoplasm
6. Nuclear envelope
7. Nucleolus
8. Centromere

Figure 1.28 *Hyacinthus* sp. root tip immediately behind the meristem, showing stages of mitosis (see fig. 1.29).

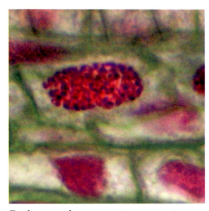

Early prophase — Chromatin begins to condense to form chromosomes.

Late prophase — Nuclear envelope intact, and chromatin condensing into chromosomes.

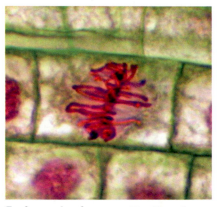

Early metaphase — Duplicated chromosomes, each made up of two chromatids, at equatorial plane.

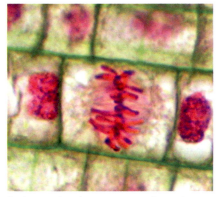

Late metaphase — Duplicated chromosomes, each made up of two chromatids, at equatorial plane.

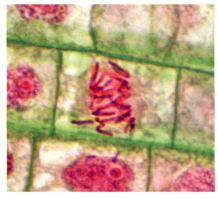

Early anaphase — Sister chromatids beginning to separate into daughter chromosomes.

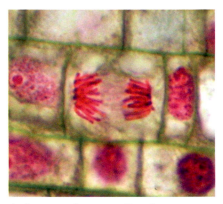

Late anaphase — Daughter chromosomes nearing poles.

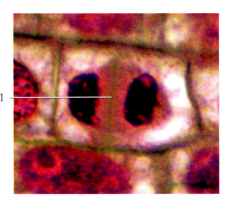

Telophase — Daughter chromosomes at poles, cell plate forming.
  1. Cell plate

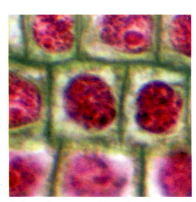

Interphase — Two daughter cells result from cytokinesis.

Figure 1.29 Stages of mitosis in Hyacinth, *Hyacinthus*, root tip. (all 430X)

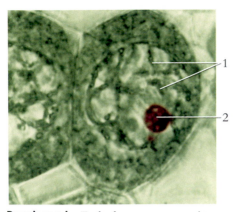

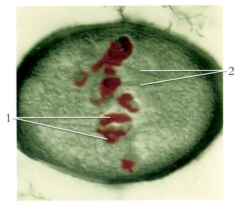

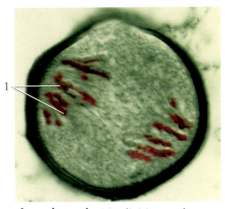

Prophase I—Each chromosome consists of two chromatids joined by a centromere.
 1. Chromatids
 2. Nucleolus

Metaphase I—Chromosome pairs align at the equator.
 1. Chromosome pairs at equator
 2. Spindle fibers

Anaphase I—No division at the centromeres occurs as the chromosomes separate, so one entire chromosome goes to each pole.
 1. Chromosomes (two chromatids)

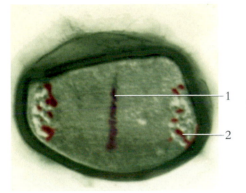

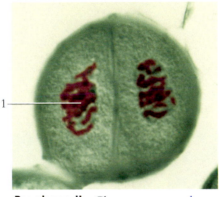

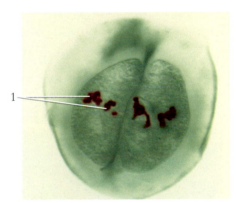

Telophase I—Chromosomes lengthen and become less distinct. The cell plate (in some plants) forms between forming cells.
 1. Cell plate (new cell wall)
 2. Chromosome

Prophase II—Chromosomes condense as in prophase I.
 1. Chromosomes (two chromatids)

Metaphase II—Chromosomes align on the equator and spindle fibers attach to the centromeres. This is similar to metaphase in mitosis.
 1. Chromosomes

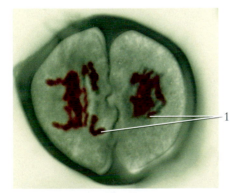

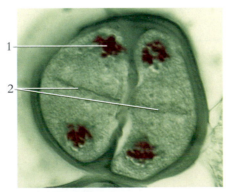

Anaphase II—Chromatids separate and each is pulled to an opposite pole.
 1. Chromatids

Telophase II—Cell division is complete and cell walls of four haploid cells are formed.
 1. Chromatids
 2. New cell walls (cell plates)

Figure 1.30 Stages of meiosis in lily microsporocytes to form microspores. (all 1000X)

Chapter 2
Prokaryotes

Prokaryotes range between 1 and 50 μm in width or diameter. The morphological appearance of bacteria may be spiral (spirillum), spherical (coccus), or rod-shaped (bacillus). Cocci and bacilli frequently form clusters or linear filaments, and may have bacterial flagella. Relatively few species of bacteria cause infection. Hundreds of species of non-pathogenic bacteria live on the human body and within the gastrointestinal (GI) tract. Those in the GI tract constitute a person's gut fauna, and are biologically critical to humans.

Photosynthetic bacteria contain chlorophyll and release oxygen during photosynthesis. Some bacteria are *obligate aerobes* (require O_2 for metabolism) and others are *facultative anaerobes* (indifferent to O_2 for metabolism). Some are *obligate anaerobes* (oxygen may poison them). Most bacteria are heterotrophic *saprophytes*, which secrete enzymes to break down surrounding organic molecules into absorbable compounds.

Archaea are adapted to a limited range of extreme conditions. The cell walls of Archaea contain pseudomurein and lack peptidoglycan (characteristic of bacteria). Archaea have distinctive RNAs and RNA polymerase enzymes. They include methanogens, typically found in swamps and marshes, and thermoacidophiles, found in acid hot springs, acidic soil, and deep oceanic volcanic vents.

Methanogens exist in oxygen-free environments and subsist on simple compounds such as CO_2, acetate, or methanol. As their name implies, Methanogens produce methane gas as a byproduct of metabolism. These organisms are typically found in organic-rich mud and sludge that often contains fecal wastes.

Thermoacidophiles are resistant to hot temperatures and high acid concentrations. The cell membrane of these organisms contains high amounts of saturated fats, and their enzymes and other proteins are able to withstand extreme conditions without denaturation. These microscopic organisms thrive in most hot springs and hot, acid soils.

The oldest known fossil cells are prokaryotic and many formed deposits called stromatolites.

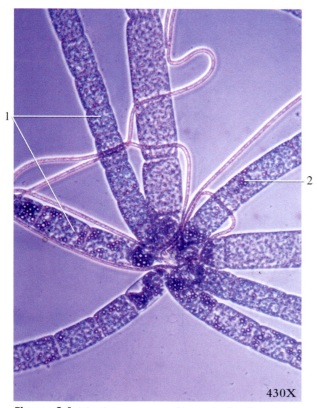

Figure 2.1 *Thiothrix*, a genus of bacteria that forms sulfur granules in its cytoplasm. These organisms obtain energy from oxidation of H_2S.
1. Filaments   2. Sulfur granules

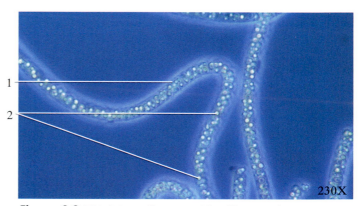

Figure 2.2 *Thiothrix* sp. filament with sulfur granules in its cytoplasm.
1. Filament   2. Sulfur granules

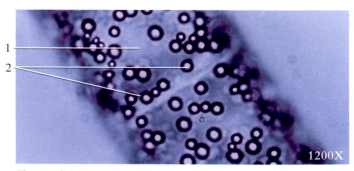

Figure 2.3 Magnified *Thiothrix* sp. filament with sulfur granules in its cytoplasm.
1. Cytoplasm   2. Sulfur granules

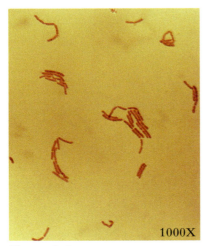

Figure 2.4 *Bacillus megaterium.* *Bacillus* is a bacterium capable of producing endospores. This species of *Bacillus* generally remains in chains after it divides.

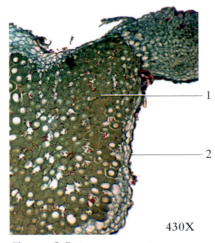

Figure 2.5 Transverse section through the root of a clover stem showing intracellular nitrogen-fixing bacteria.
1. Cell with bacteria
2. Epidermis

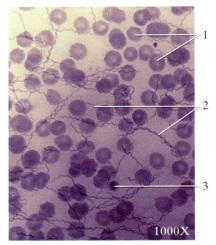

Figure 2.6 Spirochete, *Borella recurrentis.* Spirochetes are flexible rods twisted into helical shapes. This species causes relapsing fever.
1. Red blood cells
2. Spirochete
3. White blood cell

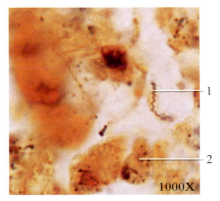

Figure 2.7 *Treponema pallidum* is a spirochete that causes syphilis.
1. *Treponema pallidum*
2. White blood cell

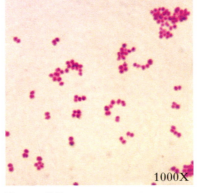

Figure 2.8 *Neisseria gonorrhoeae* is a diplococcus that causes gonorrhea.

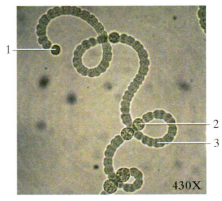

Figure 2.9 *Anabaena* sp. filament. This organism is a nitrogen-fixing cyanobacterium. Nitrogen fixation takes place within the heterocyst cells.
1. Heterocyst   3. Vegetative
2. Spore (akinete)   cell

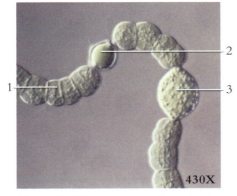

Figure 2.10 *Anabaena* sp. filament. This is a nitrogen-fixing cyanobacterium. Nitrogen fixation takes place within the heterocyst cells.
1. Vegetative cell   3. Spore
2. Heterocyst

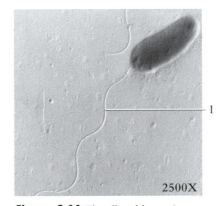

Figure 2.11 Flagellated bacterium, *Pseudomonas* sp.
1. Flagellum

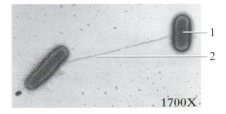

Figure 2.12 Conjugation of the bacterium *Escherichia coli.* By this process of conjugation, genetic material is transferred through the conjugation tube from one cell to the other allowing genetic recombination.
1. Bacterium   2. Conjugation tube

Table 2.1 Some Representatives of Bacteria and Archaea

Categories	Representative Genera
Bacteria	
Photosynthetic bacteria	
Cyanobacteria	*Anabaena, Oscillatoria, Spirulina, Nostoc*
Green bacteria	*Chlorobium*
Purple bacteria	*Rhodospirillum*
Gram-negative bacteria	*Proteus, Pseudomonas, Escherichia, Rhizobium, Neisseria*
Gram-positive bacteria	*Bacillus, Staphylococcus, Streptococcus, Clostridium, Listeria*
Spirochetes	*Spirochaeta, Treponema*
Actinomycetes	*Streptomyces*
Rickettsias and Chlamydias	*Rickettsia, Chlamydia*
Mycoplasmas	*Mycoplasma*
Archaea	
Methanogens	*Halobacterium, Methanobacteria*
Thermoacidophiles	*Thermoplasma, Sulfolobus*

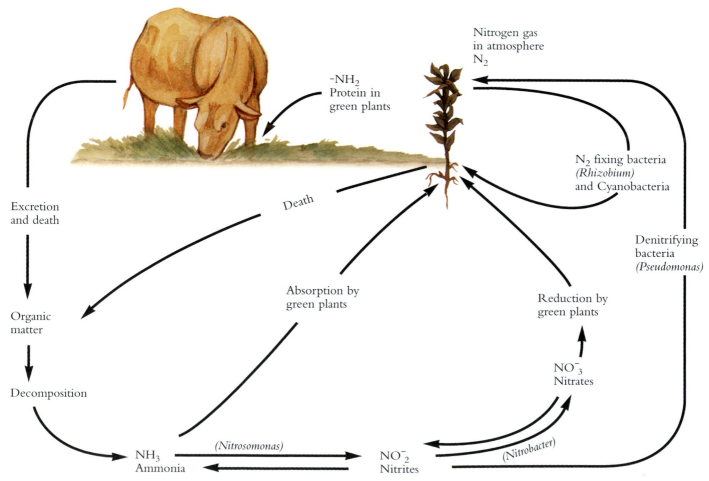

Figure 2.13 Few organisms have the ability to utilize atmospheric nitrogen. Nitrogen-fixing bacteria within the root nodules of legumes (and some free living bacteria), provide a usable source of nitrogen to plants.

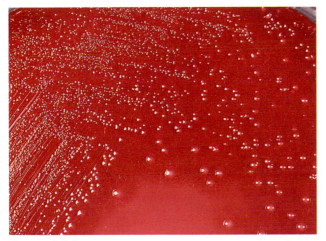

Figure 2.14 Colonies of *Streptococcus pyogenes* cultured on a nutrient agar plate. *S. pyogenes* causes strep throat and rheumatic fever in humans. This agar plate is approximately 10 cm in diameter.

Figure 2.15 Cyanobacteria live in hot springs and hot streams, such as this 40 meter effluent from a geyser in Yellowstone National Park.
  1. Mats of *Cyanophyta*

Figure 2.16 Cyanobacteria of several species growing in the effluent from a geyser. The different species are temperature dependent and form the bands of color.

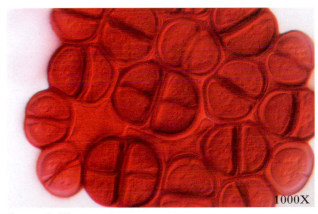

1000X

Figure 2.17 Magnified view of the cyanobacterium *Chroococcus* sp. shown with a red biological stain.

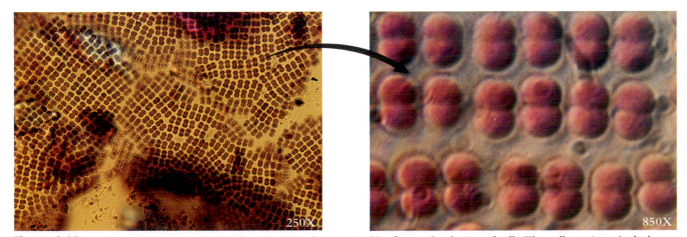

250X

850X

Figure 2.18 *Merismopedia*, a genus of cyanobacterium, is characterized by flattened colonies of cells. The cells are in a single-layer usually aligned into groups of two or four.

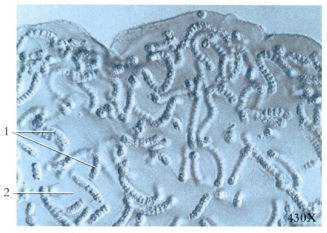

Figure 2.19 Colony of *Nostoc* sp. filaments. Individual filaments secrete mucilage, which forms a gelatinous matrix around all filaments.
1. Filaments   2. Gelatinous matrix

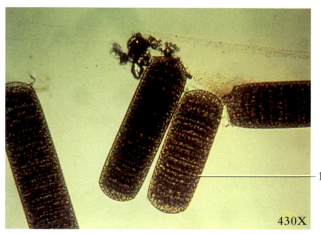

Figure 2.20 *Oscillatoria* sp. filaments. The only way this cyanobacterium can reproduce is through fragmentation of a filament. Fragments are known as hormogonia.
1. Hormogonium

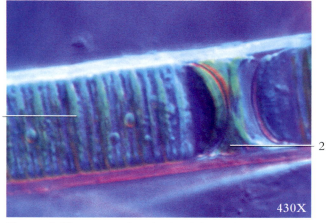

Figure 2.21 Portion of a cylindrical filament of *Oscillatoria* sp. This cyanobacterium is common in most aquatic habitats.
1. Filament segment   2. Separation disk (necridium)
   (hormogonium)

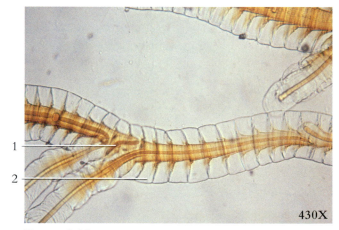

Figure 2.22 *Scytonema* sp., a cyanobacterium, is common on moistened soil. Notice the falsely-branched filament typical of this genus. This species also demonstrates "winged" sheaths.
1. False branching   2. "Winged" sheath

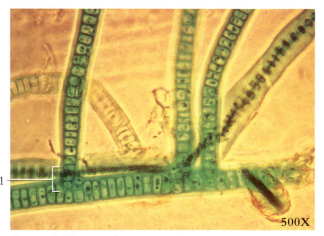

Figure 2.23 *Stigonema* sp., a cyanobacterium, has true-branched filaments caused from cell division in two separate planes.
1. True branching

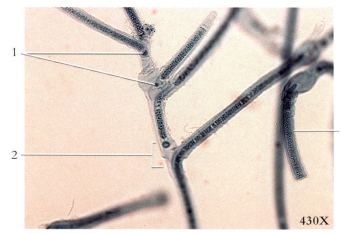

Figure 2.24 *Tolypothrix* sp., a cyanobacterium with a single false-branched filament.
1. Heterocysts   2. False branching   3. Sheath

Figure 2.25 Longitudinal section of a fossilized stromatolite two billion years old. Layering indicates the communities of bacteria and cyanobacteria mixed with sediments. This specimen originates from Australia (scale in mm).

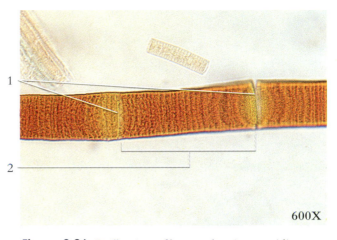

Figure 2.26 *Oscillatoria* sp. filament showing necridia.
1. Necridia   2. Hormogonium

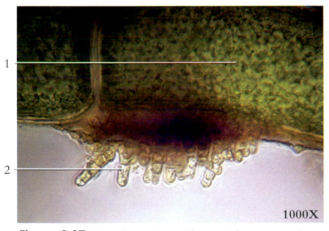

Figure 2.27 Cyanobacterium, *Chamaesisphon* sp., growing as an epiphyte on green algae, *Cladophora* sp.
1. *Cladophora* sp.   2. *Chamaesisphon* sp.

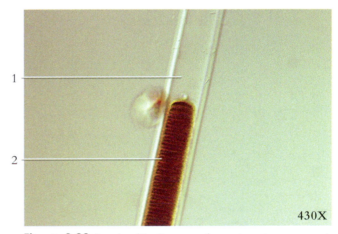

Figure 2.28 *Lyngbya birgeii*, a cyanobacterium, is common in eutrophic water throughout North America.
1. Extended sheath   2. Filament of living cells

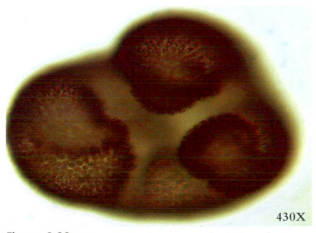

Figure 2.29 *Microcystis aeruginosa*, a cyanobacterium, that can cause toxic water "blooms".

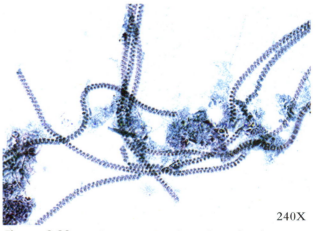

Figure 2.30 *Spirulina* sp., a cyanobacterium, showing charactristic spiral trichomes.

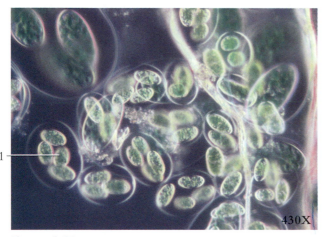

Figure 2.31 *Glaucocystis* sp. is a green alga with cyanobacteria as endosymbionts.
1. Cyanobacteria endosymbiont

Figure 2.32 *Microcoleus* sp. is one of the most common cyanobacteria in and on soils throughout the world. It is characterized by several filaments in a common sheath.

Figure 2.33 Satellite image of a large lake. The circular pattern in the water is comprised of dense growths of cyanobacteria.

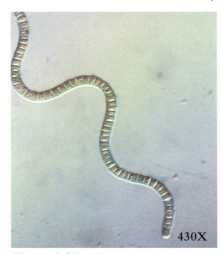

Figure 2.34 *Arthrospira* sp., a common cyanobacterium.

Figure 2.35 *Aphanizomenon* sp., a cyanobacterium common in nutrient-rich (often polluted) waters around the world.
1. Spore (akinete)   2. Filament

Figure 2.36 Spring seep in Zion National Park, Utah.
1. Mat of cyanobacteria.

Figure 2.37 Researcher examining cyanobacterial growths on soil in Canyonlands National Park, Utah.

Figure 2.38 Close up photo of cryptobiotic soil crust. These crusts are comprised of cyanobacteria, fungi, lichens, and other organisms.

Protists are eukaryotic organisms that range in size from microscopic unicellular organisms to multicellular giant kelp. Protists are all comprised of eukaryotic cells and therefore have a nucleus, mitochondria, endoplasmic reticulae, and Golgi complexes. Some contain chloroplasts. Most protists are capable of meiosis and sexual reproduction; these processes evolved a billion or more years ago and occur in nearly all complex plants and animals.

Protists are abundant in aquatic habitats, and are important constituents of plankton. Plankton are communities of organisms that drift passively or swim slowly in ponds, lakes, and oceans. Plankton are a major source of food for other aquatic organisms. Photosynthetic protists are the primary food producers in aquatic ecosystems.

The unicellular algal protists include microscopic aquatic organisms within the phyla Chrysophyta and Dinophyta. Chrysophyta are the yellow-green and golden-brown algae, and the diatoms. The cell wall of a diatom is composed largely of silica rather than cellulose. Some diatoms move in a slow,

gliding way as cytoplasm glides through slits in the cell wall to propel the organism.

The Dinophyta are single-celled, algae-like organisms, the most important of which are the dinoflagellates. In most species of dinoflagellates, the cell wall is formed of armor-like plates of cellulose. Dinoflagellates are motile, having two flagella. Generally one encircles the organism in a transverse groove, and the other projects to the posterior.

Protozoa are also protists. They are small ($2\mu m - 100\mu m$), unicellular eukaryotic organisms that lack a cell wall. Movement of protozoa is either lacking or due to flagella, cilia, or pseudopodia of various sorts. In feeding upon other organisms or organic particles, they use simple diffusion, pinocytosis, active transport, or phagocytosis. Although most protozoa reproduce asexually, some species may also reproduce sexually during a portion of their life cycle. Most protozoa are harmless, although some are of immense clinical concern because they are parasitic and may cause human disease, including African sleeping sickness and malaria.

Table 3.1 Some Representatives of the Protista: Primarily Unicellular Organisms

Taxa and Representative Kinds	Characteristics
Chrysophyta—diatoms and golden algae	Diatom cell walls composed of or impregnated with silica, often with two halves; plastids often golden in Chrysophyta due to pigment composition
Dinophyta—dinoflagellates	Two flagella in grooves of wall; brownish-gold plastids
Rhizopoda—amoebas	Cytoskeleton of microtubules and microfilaments; amoeboid locomotion
Apicomplexa—sporozoa and plasmodia	Lack locomotor capabilities and contractile vacuoles; mostly parasitic
Sarcomastigophora—protozoa	Use flagella or pseudopodia to locomote; mostly parasitic
Euglenophyta—euglenoids	Flagellates containing chloroplasts, lacking typical cell walls
Ciliophora—ciliates and *Paramecium*	Use cilia to move and feed

Plant-like (Chrysophyta, Dinophyta)

Animal-like (Rhizopoda, Apicomplexa, Sarcomastigophora, Euglenophyta, Ciliophora)

Figure 3.1 Illustration of *Amoeba proteus*, a fresh-water protozoan.

Phylum Chrysophyta - diatoms and golden algae

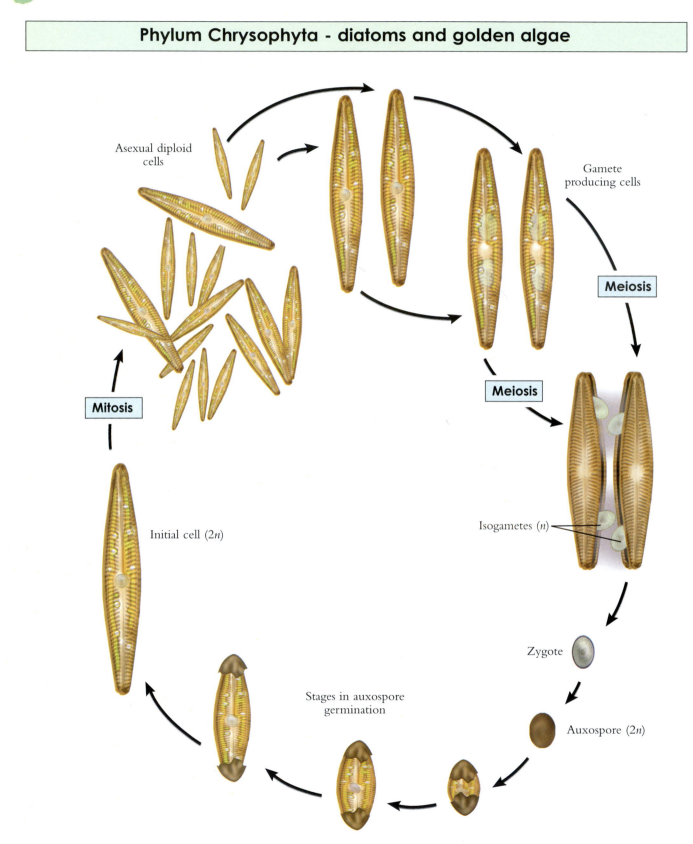

Asexual diploid cells

Gamete producing cells

Meiosis

Meiosis

Mitosis

Isogametes (*n*)

Initial cell (2*n*)

Zygote

Stages in auxospore germination

Auxospore (2*n*)

Figure 3.2 Life cycle of a pennate diatom.

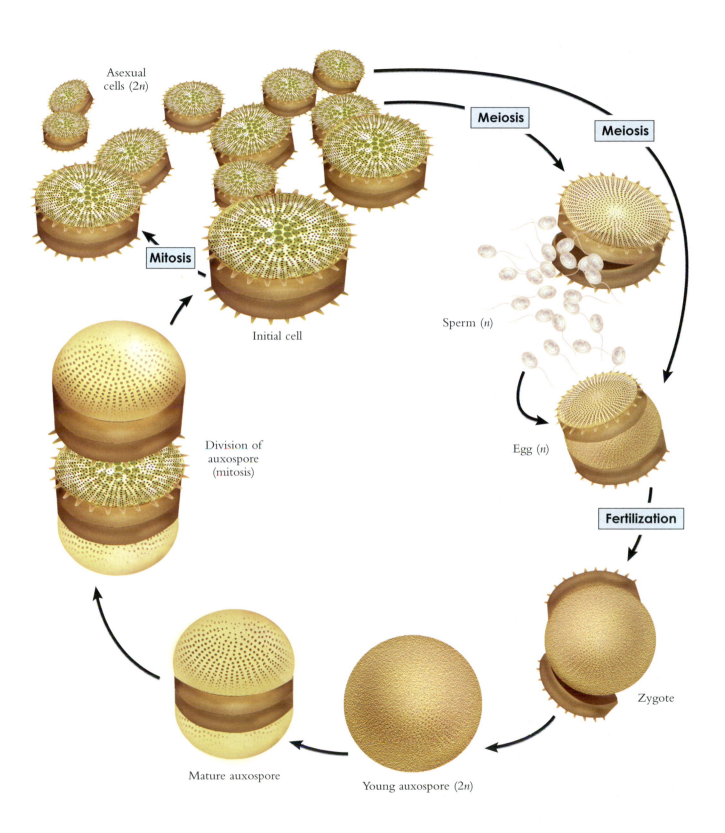

Asexual cells (2*n*)

Meiosis

Meiosis

Sperm (*n*)

Mitosis

Initial cell

Egg (*n*)

Division of
auxospore
(mitosis)

Fertilization

Zygote

Mature auxospore

Young auxospore (2*n*)

Figure 3.3 Life cycle of a centric diatom.

Figure 3.4 *Biddulphia* sp., a colony forming colonies. These cells are beginning cell division.

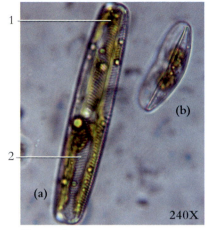

Figure 3.5 Live specimens of pennate (bilaterally symmetrical) diatoms. (a) *Navicula* sp., and (b) *Cymbella* sp.
  1. Chloroplast   2. Striae

Figure 3.6 *Hyalodiscus* sp., a centric (radially symmetrical) diatom, from a freshwater spring in Nevada.
  1. Silica cell wall   2. Chloroplasts

Figure 3.7 *Epithemia* sp., a distinctive pennate freshwater diatom.

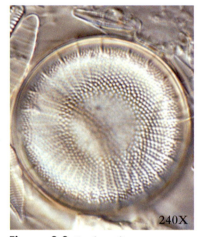

Figure 3.8 *Stephanodiscus* sp., a centric diatom.

Figure 3.9 Two common freshwater diatoms.
  1. *Cocconeis* sp.
  2. *Amphora* sp.

Figure 3.10 *Hantzschia* sp., one of the most common soil diatoms.

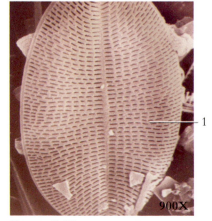

Figure 3.11 Scanning electron micrograph of *Cocconeis* sp., a common freshwater diatom.
  1. Striae containing pores, or punctae, in the frustule (silicon cell wall).

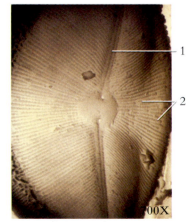

Figure 3.12 A scanning electron micrograph of the diatom *Achnanthes flexella*.
  1. Raphe   2. Striae

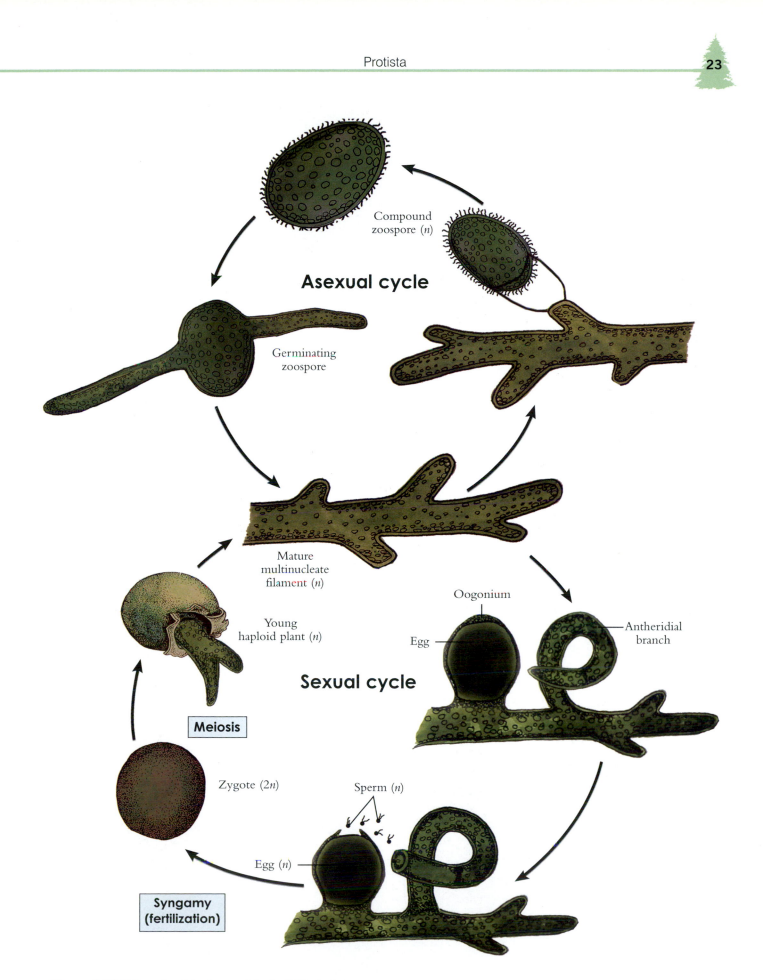

Compound
zoospore (*n*)

Asexual cycle

Germinating
zoospore

Mature
multinucleate
filament (*n*)

Young
haploid plant (*n*)

Oogonium

Egg

Antheridial
branch

Sexual cycle

Meiosis

Zygote (2*n*)

Sperm (*n*)

Egg (*n*)

**Syngamy
(fertilization)**

Figure 3.13 Life cycle of the "water felt alga," *Vaucheria* sp.

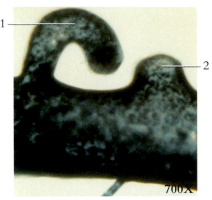

Figure 3.14 A filament with immature gametangia of the "water felt" alga, *Vaucheria* sp. *Vaucheria* is a chrysophyte that is widespread in fresh-water and marine habitats. It is also found in the mud of brackish areas that periodically become submerged and then exposed to air.

1. Antheridium
2. Developing oogonium

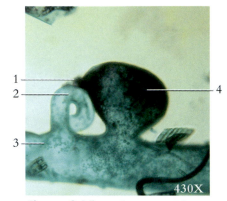

Figure 3.15 *Vaucheria* sp., with mature gametangia.

1. Fertilization pore
2. Antheridium
3. Chloroplasts
4. Developing oogonium

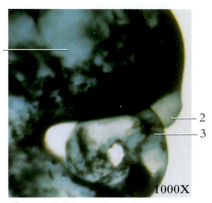

Figure 3.16 *Vaucheria* sp., with mature gametangia.

1. Oogonium
2. Fertilization pore
3. Antheridium

Phylum Dinophyta - dinoflagellates

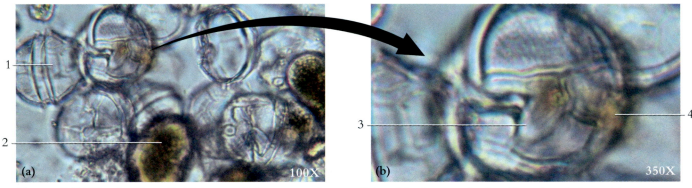

Figure 3.17 Dinoflagellates, *Peridinium* sp. (a) Some organisms are living; (b) others are dead and have lost their cytoplasm and consist of resistant cell walls.

1. Dead dinoflagellate   2. Living dinoflagellate   3. Cellulose plate   4. Remnant of cytoplasm

Figure 3.18 Giant clam with bluish coloration due to endosymbiont dinoflagellates.

Figure 3.19 Photomicrograph of *Peridinium* sp. The cell wall of many dinoflagellates is composed of overlapping plates of cellulose.

1. Transverse groove
2. Wall of cellulose plates

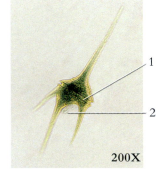

Figure 3.20 *Ceratium* sp., a common fresh water dinoflagellate.

1. Transverse groove
2. Trailing flagellum

Phylum Rhizopoda - amoebas

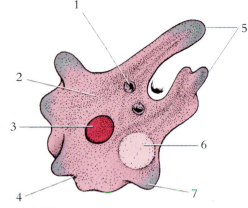

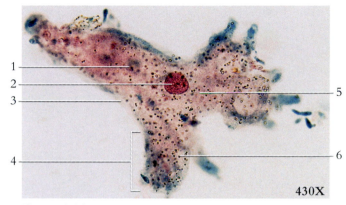

Figure 3.21 *Amoeba proteus,* is a fresh-water protozoan that moves by forming cytoplasmic extensions called pseudopodia.

1. Food vacuole   4. Cell membrane   7. Ectoplasm
2. Endoplasm   5. Pseudopodia
3. Nucleus   6. Contractile
                      vacuole

Figure 3.22 *Amoeba proteus.*

1. Food vacuole   3. Cell membrane   5. Endoplasm
2. Nucleus   4. Pseudopodia   6. Ectoplasm

430X

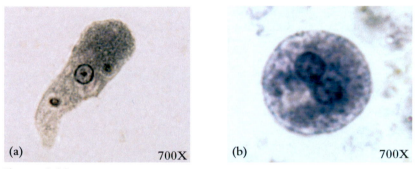

(a) 700X   (b) 700X

Figure 3.23 Protozoan *Entamoeba histolytica,* is the causative agent of amebic dysentery, a disease most common in areas with poor sanitation. (a) A trophozoite, and (b) a cyst.

Phylum Apicomplexa - sporozoans and plasmodium

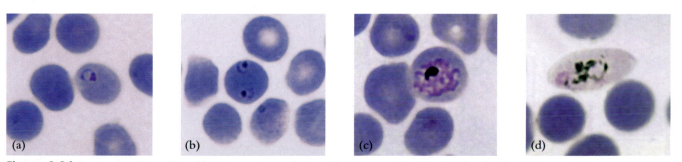

(a)   (b)   (c)   (d)

Figure 3.24 Protozoan *Plasmodium falciparum* causes malaria, which is transmitted by the female *Anopheles* mosquito. (a) The ring stage in a red blood cell, (b) a double infection, (c) a developing schizont, and (d) a gametocyte.

Phylum Sarcomastigophora - flagellated protozoans

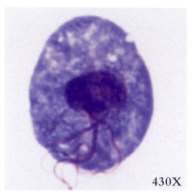

430X

Figure 3.25 Protozoan *Trichomonas vaginalis* is the causative agent of trichomoniasis. Trichomoniasis is an inflammation of the genitourinary mucosal surfaces—the urethra, vulva, vagina, and cervix in females and the urethra, prostate, and seminal vesicles in males.

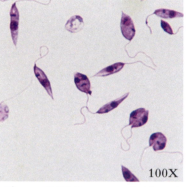

100X

Figure 3.26 Protozoan *Leishmania donovani* is the causative agent of visceral leishmaniasis, or kala-azar disease, in humans. The sandfly is the infectious host of this disease.

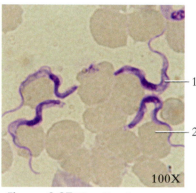

100X

Figure 3.27 Flagellated protozoan *Trypanosoma brucei* is the causative agent of African trypanosomiasis, or African sleeping sickness. The tsetse fly is the infectious host of this disease in humans.
1. *Trypanosoma brucei*
2. Red blood cell

Phylum Euglenophyta - euglenoids

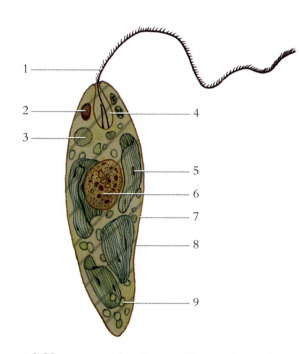

Figure 3.28 Diagram of *Euglena*, which contains flagellates that contain chloroplasts. They are freshwater organisms that have a flexible pellicle rather than a rigid cell wall.

1. Long flagellum
2. Photoreceptor (eye spot)
3. Contractile vacuole
4. Reservoir
5. Chloroplast
6. Nucleus
7. Pellicle
8. Cell membrane
9. Paramylon granule

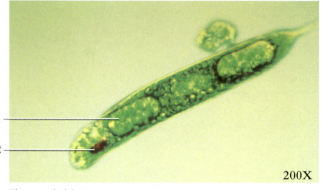

200X

Figure 3.29 Species of *Euglena*.
1. Paramylon body   2. Photoreceptor

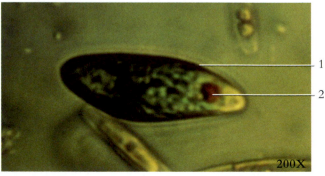

200X

Figure 3.30 Species of *Euglena* from a brackish lake in New Mexico.
1. Pellicle   2. Photoreceptor

Phylum Ciliophora - ciliates and paramecia

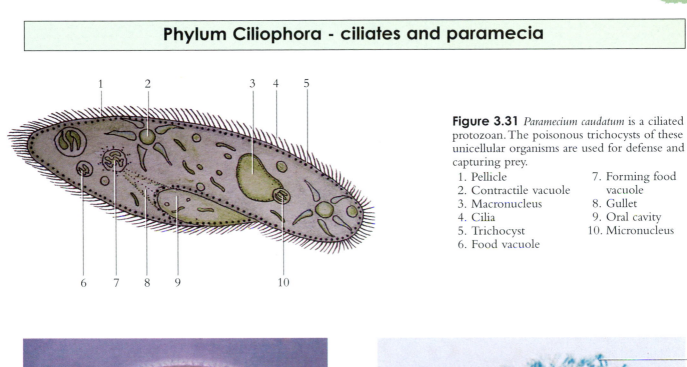

Figure 3.31 *Paramecium caudatum* is a ciliated protozoan. The poisonous trichocysts of these unicellular organisms are used for defense and capturing prey.

1. Pellicle
2. Contractile vacuole
3. Macronucleus
4. Cilia
5. Trichocyst
6. Food vacuole
7. Forming food vacuole
8. Gullet
9. Oral cavity
10. Micronucleus

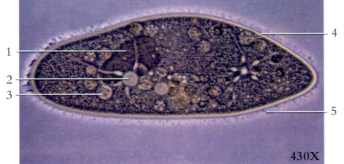

430X

Figure 3.32 *Paramecium caudatum,* a ciliated protozoan.

1. Macronucleus
2. Contractile vacuole
3. Micronucleus
4. Pellicle
5. Cilia

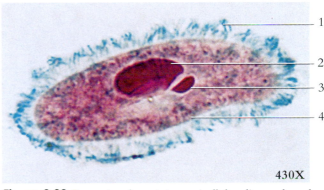

430X

Figure 3.33 *Paramecium bursaria* is a unicellular, slipper-shaped organism. Paramecia are usually common in ponds containing decaying organic matter.

1. Cilia
2. Macronucleus
3. Micronucleus
4. Pellicle

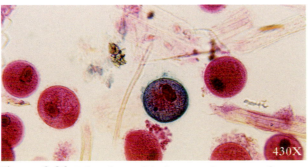

430X

Figure 3.34 *Balantidium coli* is the causative agent of balantidiasis. Cysts in sewage-contaminated water are the infective form.

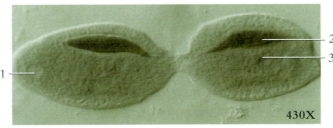

430X

Figure 3.35 *Paramecium* sp. in fission.

1. Contractile vacuole
2. Macronucleus
3. Micronucleus

Chapter 4
Algae, Slime Molds, and Water Molds

Protists live in nearly all Earth's habitats, especially in areas that are damp or aquatic. Many symbiont protists inhabit the host's body cells, tissues, or fluids. Some parasitic protists are pathogens to plants and animals. Most are aerobic, using mitochondria for cellular respiration. Some protists have chloroplasts and are photoautrophs. Others are heterotrophs, absorbing or ingesting organic molecules.

Although protists vary in their mode of reproduction, all can reproduce asexually. Some only reproduce asexually, while others may reproduce sexually as well. Certain protists endure harsh conditions by forming protective cysts during a portion of their life cycle.

The three phyla of primarily multicellular algae are Chlorophyta (green algae), Phaeophyta (brown algae), and Rhodophyta (red algae). Most species of these algae are multicellular, aquatic organisms. For example, seaweeds (phaeophytes) are multicellular, mostly marine, brown algae. Included in this phylum are giant kelp that may exceed 100 meters in length. Consisting of cellulose and algin, the cell walls of brown algae can withstand the movements of ocean currents and waves. These substances give seaweeds their characteristic slimy and rubbery feel.

Evidence indicates that Chlorophyta are ancestral to plants. Most green algae live in fresh water, although marine planktonic and attached forms exist. Chlorophytes are *photo-autotrophic*, manufacturing their own food. *Lichens* are fungi that live symbiotically with algae forming a single organism.

Most red algae are multicellular, marine forms. Colors other than red among Rhodophyta are not uncommon. Similar to brown algae, red algae are commonly called *seaweeds*. They reproduce sexually but lack flagellated stages. Alternation of generations is common. High in starch, some rhodophytes are harvested for food. Agar, used as a culture medium in bacteriology, is obtained from a species of red algae.

Organisms within the phylum Myxomycota (plasmodial slime molds), Dictyosteliomycota (cellular slime molds), and Oomycota (water mold, white rusts, downy mildews) resemble fungi; the similarities, however, are believed to be due to convergence.

Many slime molds are yellow, red, or orange. All are non-photosynthetic heterotrophs. The body of a myxomycete is a multinucleated continuum of cytoplasm undivided by membranes or walls. During the feeding stage, an amoeboid mass called a plasmodium extends through moist organic soil, leaves, or decaying logs engulfing food particles by phagocytosis. Plasmodial slime molds are decomposers in some habitats.

Dictyosteliomycota have a feeding stage consisting of solitary haploid cells, and an aggregate stage consisting of an amoeboid mass of cells. The aggregate stage is formed from thousands of individual cells that join at some signal to form a single body. Asexual fruiting bodies are produced by the aggregate stage.

Most Oomycota are saprophytes on dead plants and animals, and many are important decomposers in fresh-water ecosystems. Some Oomycota are parasitic on the skin and gills of fishes. Species of white rusts and downy mildews live on land as plant parasites. Distributed by windblown spores, these organisms are of concern in the potato industry (potato blight) and the grape and wine industry (downy mildew).

Table 4.1 Some Representatives of Protista: Primarily Multicellular Organisms

Phylum and Representative Kinds	Characteristics
Algae	
Phylum Chlorophyta—green algae	Unicellular, colonial, filamentous and multicellular plate-like forms; mostly fresh water; reproduce asexually and sexually
Phylum Phaeophyta—brown algae, giant kelp	Multicellular, mostly marine often in the intertidal zone; most with alternation of generations
Phylum Rhodophyta—red algae	Multicellular, mostly marine; sexual reproduction but with no flagellated cells; alternation of generations common
Protists Resembling Fungi	
Phylum Myxomycota—plasmodial slime molds	Multinucleated continuum of cytoplasm without cell membranes; amoeboid plasmodium during feeding stage; produce asexual fruiting bodies; gametes produced by meiosis
Phylum Dictyosteliomycota—cellular slime molds	Solitary cells during feeding stage; cells aggregate when food is scarce; produce asexual fruiting bodies
Phylum Oomycota—water molds, white rusts and downy mildews	Decomposers or parasitic forms; walls of cellulose, dispersal by non-motile spores or flagellated zoospores, gametes produced by meiosis

Phylum Chlorophyta - green algae

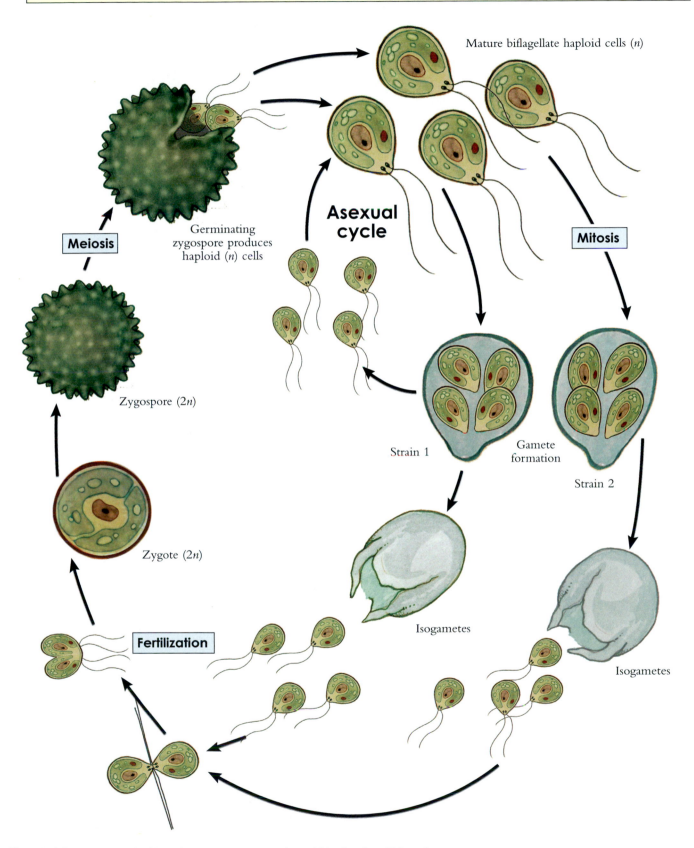

Mature biflagellate haploid cells (n)

Meiosis

Germinating
zygospore produces
haploid (n) cells

**Asexual
cycle**

Mitosis

Zygospore (2n)

Strain 1   Gamete
formation

Strain 2

Zygote (2n)

Fertilization

Isogametes

Isogametes

Figure 4.1 Life cycle of *Chlamydomonas* sp., a green alga within the class Chlorophyceae.

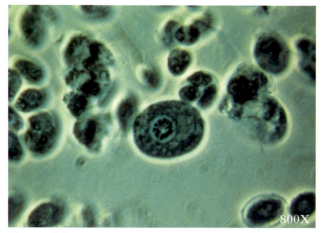

Figure 4.2 *Chlamydomonas* sp., a common unicellular green alga.

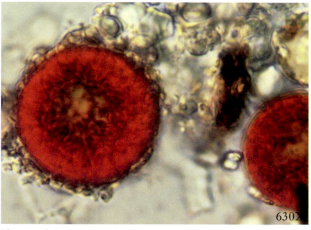

Figure 4.3 *Chlamydomonas nivalis*, the common snow alga.

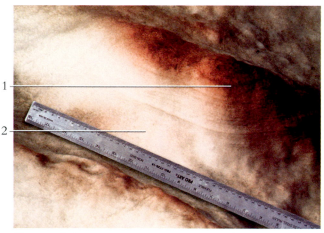

Figure 4.4 Habitat shot of *Chlamydomonas nivalis* creating "red snow."
1. *Chlamydomonas nivalis*   2. Snow

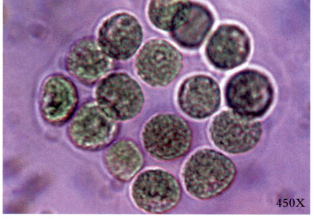

Figure 4.5 *Gonium* sp. colony. *Gonium* sp. is a 16-celled flat colony of *Chlamydomonas*-like cells.

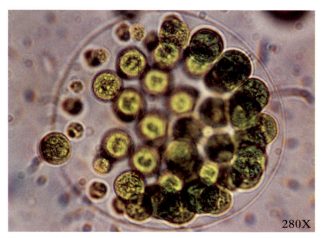

Figure 4.6 *Pleodorina* sp., a multicellular colony (often 64-celled) relative of *Chlamydomonas* and *Volvox*.

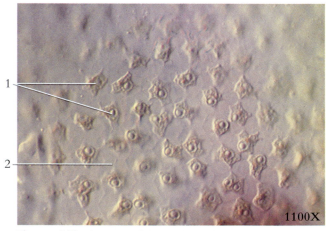

Figure 4.7 Close-up of the surface of *Volvox* sp. showing the interconnections between cells.
1. Vegetative cells   2. Cytoplasmic connection between cells

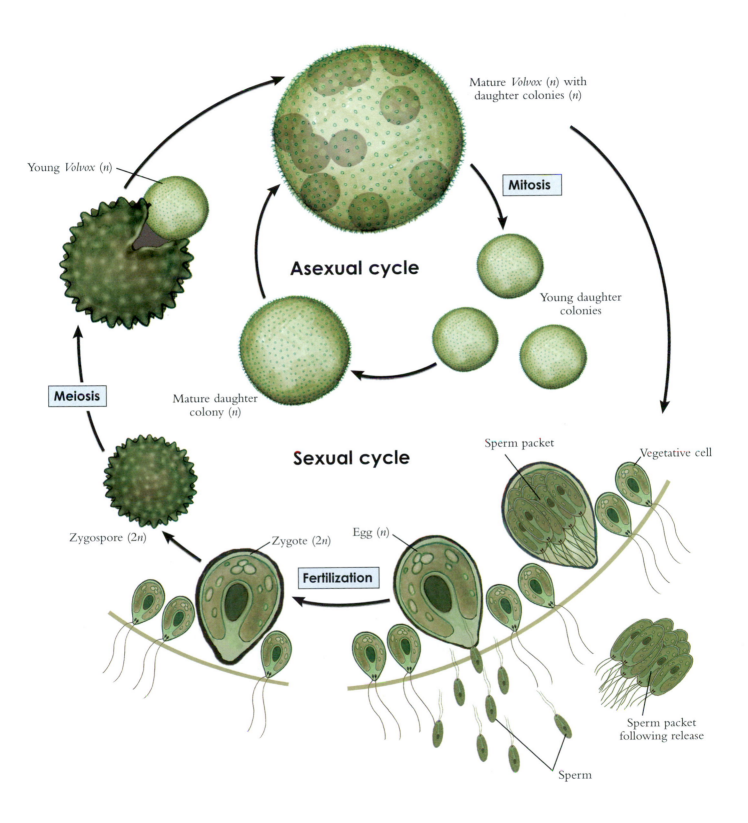

Mature *Volvox* (*n*) with
daughter colonies (*n*)

Mitosis

Young *Volvox* (*n*)

Asexual cycle

Young daughter
colonies

Meiosis

Mature daughter
colony (*n*)

Sexual cycle

Sperm packet

Vegetative cell

Zygospore (2*n*)   Zygote (2*n*)   Egg (*n*)

Fertilization

Sperm packet
following release

Sperm

Figure 4.8 Life cycle of *Volvox* sp., a common freshwater chlorophyte. *Volvox* is considered by
some to be a colony and by others to be a single, integrated alga.

Figure 4.9 *Volvox* sp. Three separate organisms are shown in this photomicrograph, each containing daughter colonies of various ages.

1. Daughter colony initial (gonidium)
2. Daughter colonies
3. Vegetative cells

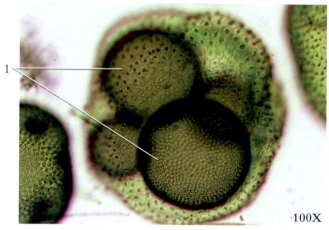

Figure 4.10 *Volvox* sp., a single organism with several large daughter colonies.

1. Daughter colonies

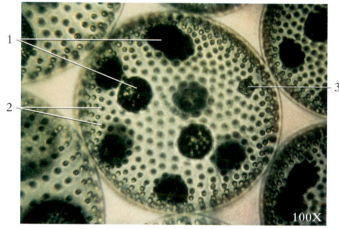

Figure 4.11 *Volvox* sp., a single mature specimen with several eggs and zygotes.

1. Zygotes   2. Vegetative cells   3. Egg

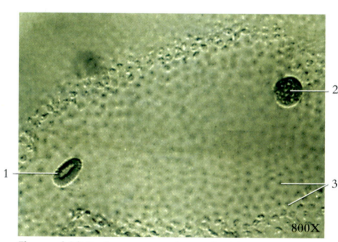

Figure 4.12 Single mature specimen of *Volvox* sp. This photomicrograph is a highly magnified view of a single organism showing gametes.

1. Sperm packet   3. Vegetative cells
2. Egg

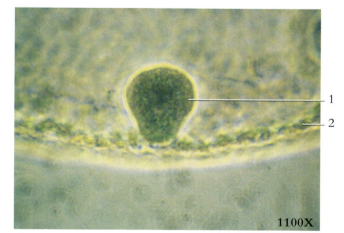

Figure 4.13 *Volvox* sp., showing a prominent egg at the edge of the organism. This egg will be fertilized to develop a zygote and then a zygospore.

1. Egg   2. Vegetative cells

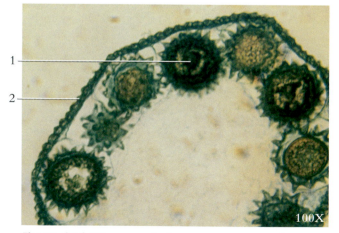

Figure 4.14 *Volvox* sp., a single mature organism with zygospores.

1. Zygospore
2. Vegetative cells

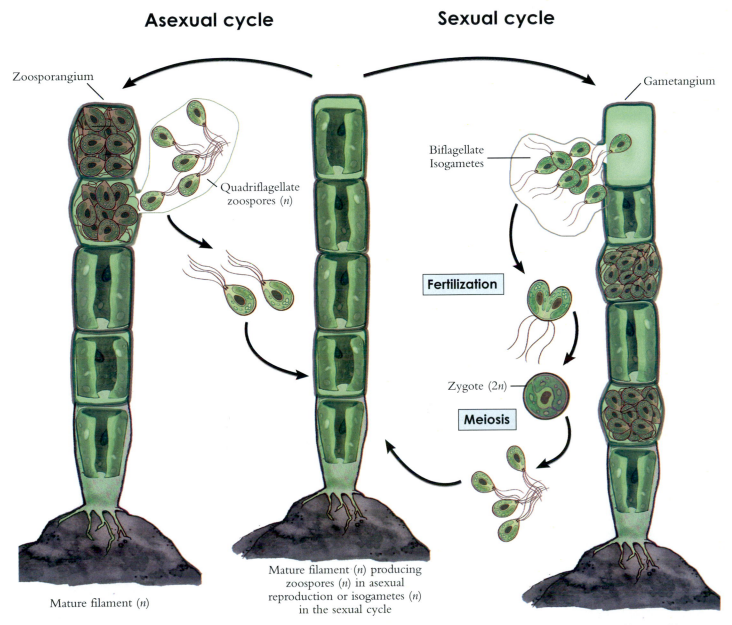

Figure 4.15 Life cycle of *Ulothrix* sp., a green alga within the class Ulvophyceae.

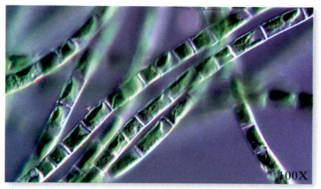

Figure 4.16 Live specimens of *Ulothrix* sp., an unbranched, filamentous green alga.

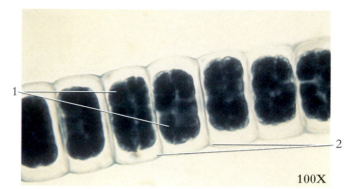

Figure 4.17 *Ulothrix* sp., an unbranched, filamentous green alga.
1. Zoospores
2. Individual cells (termed sporangia when they produce spores)

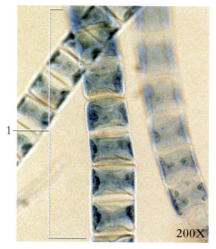

Vegetative stained filament

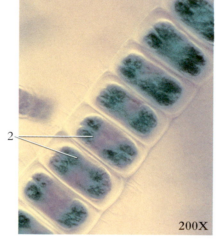

Stained filament with zoospores

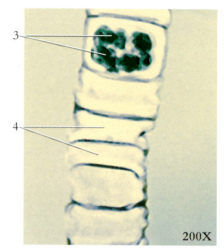

Empty filament, after zoospores have been released

Figure 4.18 Production and release of zoospores in the green alga *Ulothrix* sp.
1. Filament   2. Young zoospores   3. Mature zoospores   4. Empty cells following zoospore release

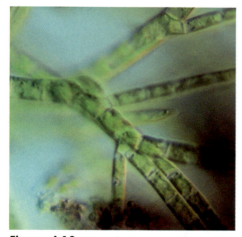

Figure 4.19 *Stigeocolonium* sp., a close relative of *Ulothrix*, showing a branched thallus.

Figure 4.20 *Draparnaldia* sp., a relative of *Ulothrix*, showing different cell sizes in the thallus and a characteristic branching pattern.

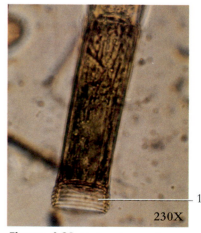

Figure 4.21 *Oedogonium* sp. has distinct "apical caps" that accrue from cell division in this genus.
1. Apical caps

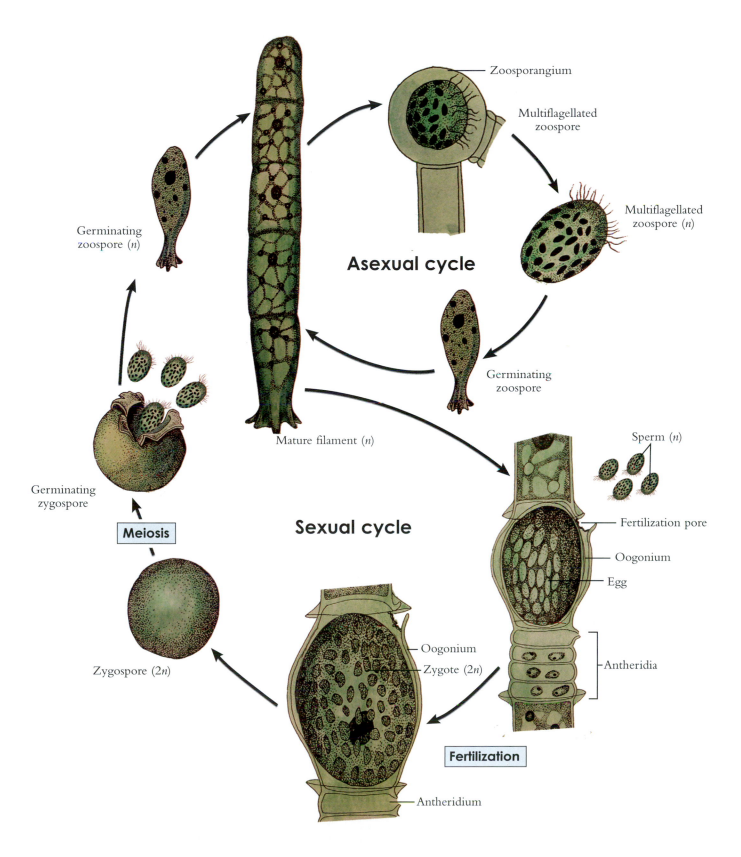

Figure 4.22 Life cycle of *Oedogonium* sp., an unbranched, filamentous green alga.

430X

Figure 4.23 Young filament of *Oedogonium* sp.
1. Basal cell
2. Holdfast

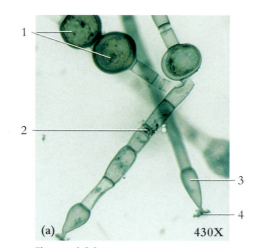

(a)   430X

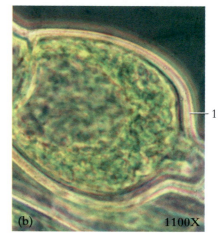

(b)   1100X

Figure 4.24 (a) *Oedogonium* sp., a filamentous, unbranched, green alga. (b) Close-up of an oogonium.
1. Oogonia   3. Basal holdfast cell
2. Antheridium   4. Holdfast

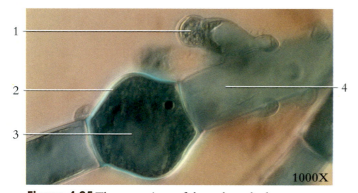

1000X

Figure 4.25 The oogonium of the unbranched, green alga, *Oedogonium* sp.
1. Dwarf male filament   3. Developing egg
2. Oogonium   4. Vegetative cell

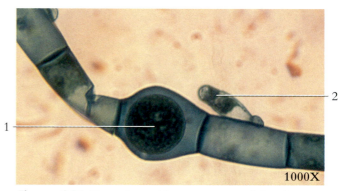

1000X

Figure 4.26 Oogonium with mature egg and dwarf male filament.
1. Egg
2. Dwarf male filament

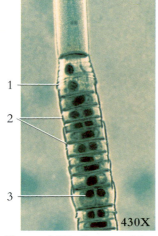

430X

Figure 4.27 Filament of the green alga, *Oedogonium* sp.
1. Annular scars from cell division
2. Antheridia
3. Sperm

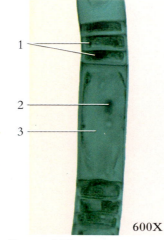

600X

Figure 4.28 Green alga, *Oedogonium* sp., showing antheridia between vegetative cells.
1. Sperm within antheridia
2. Nucleus of vegetative cell
3. Vegetative cell

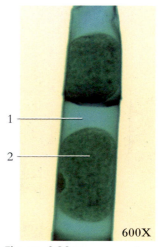

600X

Figure 4.29 Zoosporangium of the unbranched green alga, *Oedogonium* sp.
1. Zoosporangium
2. Zoospore

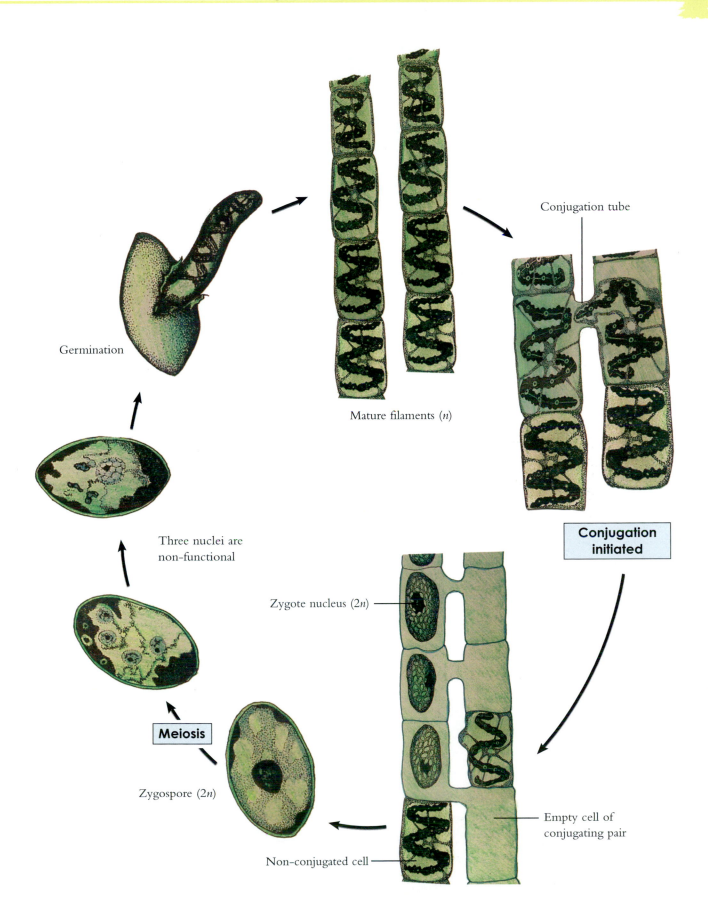

Conjugation tube

Germination

Mature filaments (*n*)

Conjugation initiated

Three nuclei are non-functional

Zygote nucleus (2*n*)

Meiosis

Zygospore (2*n*)

Empty cell of conjugating pair

Non–conjugated cell

Figure 4.30 Life cycle of *Spirogyra* sp., a common fresh-water green alga.

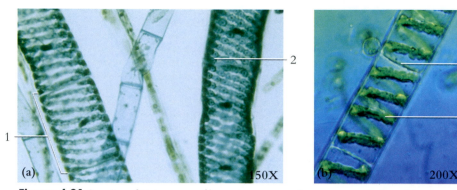
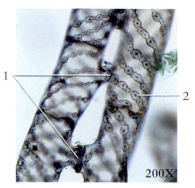

Figure 4.31 Species of *Spirogyra* are filamentous green algae commonly found in green masses on the surfaces of ponds and streams. Their chloroplasts are arranged as a spiral within the cell. (a) Several cells comprise a filament. (b) A magnified view of a single filament composed of several cells.

1. Single cell   2. Filaments   3. Cell wall   4. Chloroplast

Figure 4.32 Filaments of *Spirogyra* sp. showing initial contact of conjugation tubes.
1. Conjugation tube
2. Pyrenoid in chloroplast

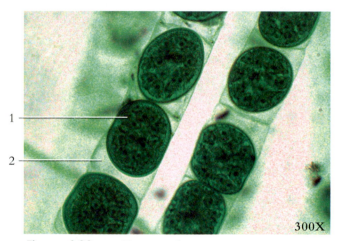

Figure 4.33 Two filaments of *Spirogyra* sp. with aplanospores.
1. Aplanospore   2. Cell wall

Figure 4.34 *Spirogyra* sp. in a small fresh-water pond.

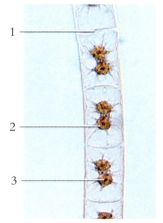
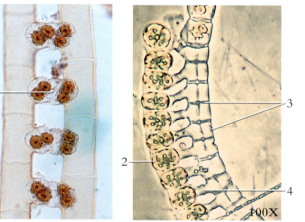

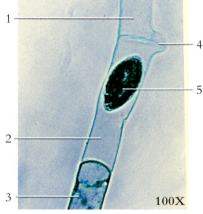

Figure 4.35 *Zygnema* sp. filament showing the star-shaped chloroplasts.
1. Cell wall
2. Chloroplast
3. Pyrenoid

Figure 4.36 *Zygnema* sp. showing two locations of fertilization, (a) in the conjugation tube and (b) in cells of one of the conjugating filaments.
1. Fusing gametes   3. Cell wall
2. Zygote   4. Conjugation tube

Figure 4.37 Self–fertile species of *Spirogyra* sp. A gamete has migrated from the upper cell to form a zygote in the lower cell.
1. Upper cell   4. Conjugation
2. Lower cell   tube
3. Chloroplast   5. Zygote

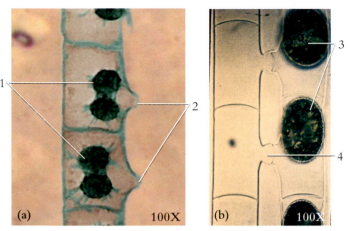

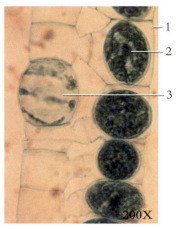

Figure 4.38 *Zygonema* sp. undergoing conjugation. (a) The filament is just forming conjugation tubes; (b) and two conjugated filaments.

1. Developing gametes   3. Zygotes
2. Developing conjugation tubes   4. Conjugation tube

Figure 4.39 Conjugation in *Spirogyra* sp.

1. Cell bearing zygote
2. Zygote
3. Cell that did not conjugate

Figure 4.40 Desmid *Closterium* sp. Desmids are unicellular, freshwater chlorophyta, which reproduce sexually by conjugation.

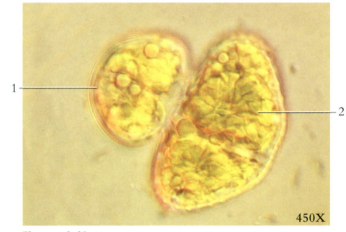

Figure 4.41 *Cosmarium* sp., a desmid, soon after cell division forming a new semi-cell.

1. New semi-cell   2. Dividing cell

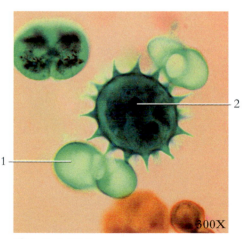

Figure 4.42 Zygospore of the desmid *Cosmarium* sp.

1. Empty cell that has been involved in conjugation
2. Zygospore

Figure 4.43 Desmid *Micrasterias* sp.

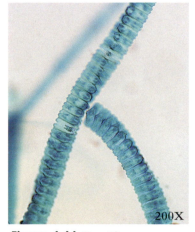

Figure 4.44 *Desmidium* sp., a filamentous (colonial) desmid.

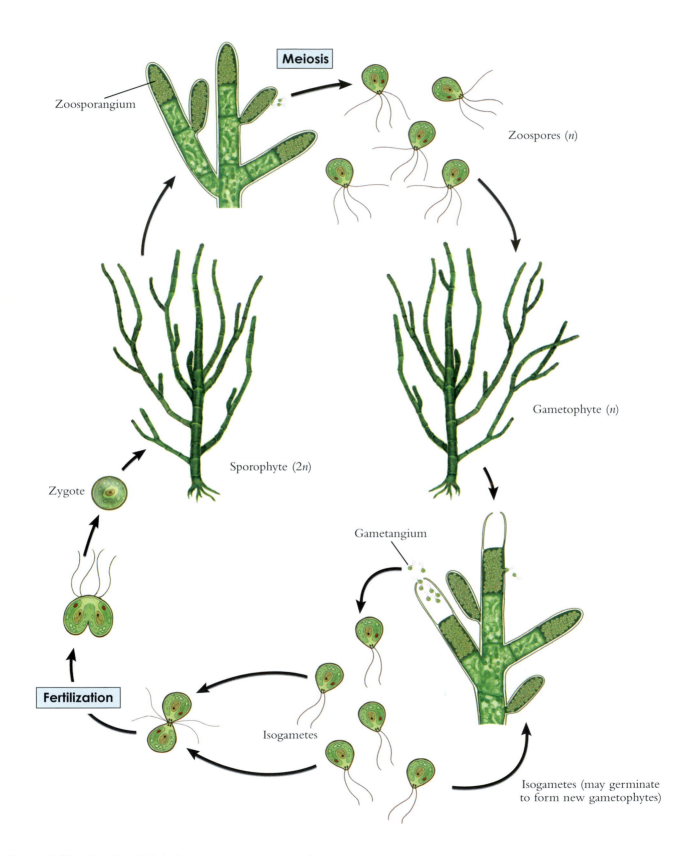

Figure 4.45 Life cycle of *Cladophora* sp. a common green alga.

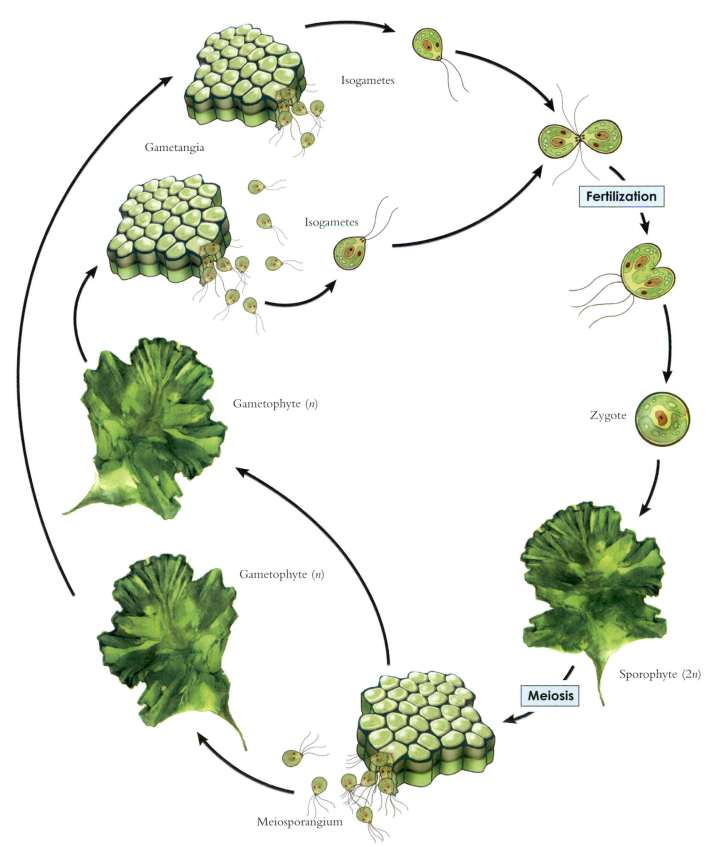

Figure 4.46 Life cycle of *Ulva*, a common marine green alga.

Isogametes

Gametangia

Isogametes

Fertilization

Gametophyte (*n*)

Zygote

Gametophyte (*n*)

Sporophyte (2*n*)

Meiosis

Meiosporangium

Figure 4.47 (a) Sea lettuce, *Ulva* sp., lives as a flat membranous chlorophyte in marine environments. (b) Detailed view.

Figure 4.48 Magnified view of the surface of *Enteromorpha intestinalis*. *Enteromorpha is* closely related to *Ulva*.

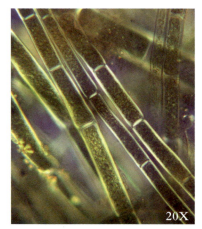

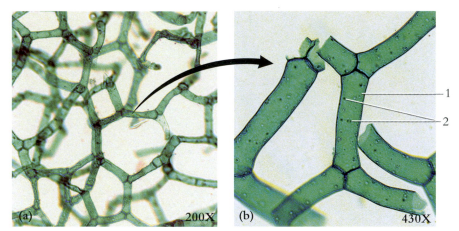

Figure 4.49 Filaments of *Cladophora* sp. This member of class Ulvophyceae is found in both fresh-water and marine habitats.

Figure 4.50 (a) *Hydrodictyon* sp. The large, multinucleated cells form net-shaped colonies. (b) A magnified view *Hydrodictyon* sp.

  1. Individual cell   2. Nuclei of cell

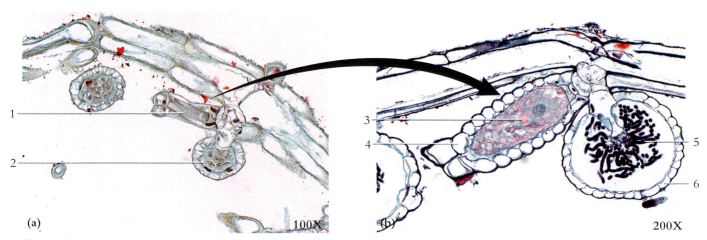

Figure 4.51 (a) *Chara* sp. inhabits marshes or shallow, temperate lakes, showing characteristic gametangia. (b) A magnified view of the gametangia.

  1. Oogonium   3. Egg   5. Sperm-producing cells (filaments)

  2. Antheridium   4. Oogonium   6. Antheridium

Phylum Phaeophyta - brown algae and giant kelp

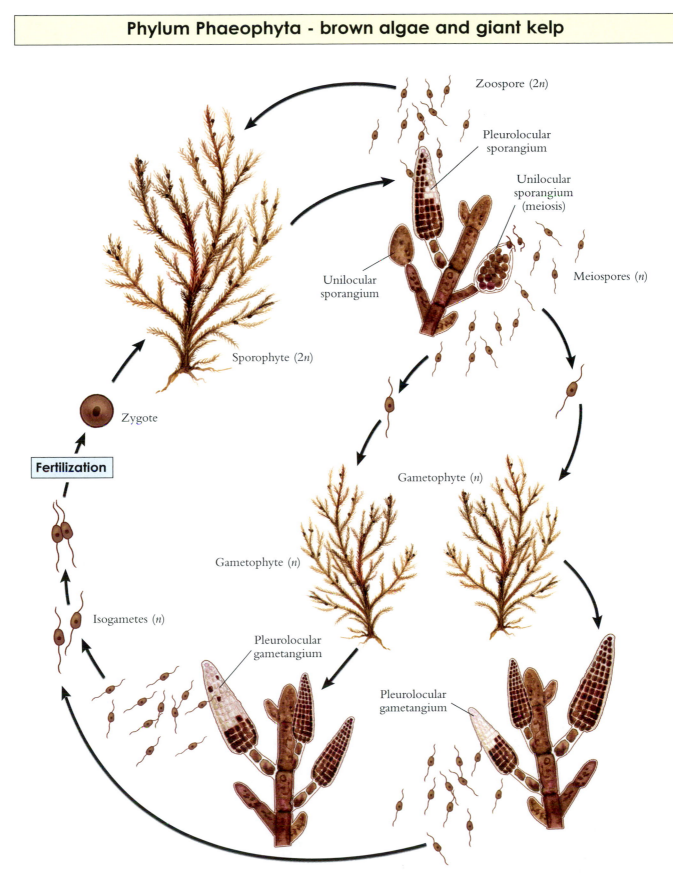

Zoospore (2*n*)

Pleurolocular
sporangium

Unilocular
sporangium
(meiosis)

Unilocular
sporangium

Meiospores (*n*)

Sporophyte (2*n*)

Zygote

Gametophyte (*n*)

Fertilization

Gametophyte (*n*)

Isogametes (*n*)

Pleurolocular
gametangium

Pleurolocular
gametangium

Figure 4.52 Life cycle of *Ectocarpus* sp., a common brown alga.

Figure 4.53 Thallus of *Ectocarpus* sp. showing branching and pleurolocular sporangia.

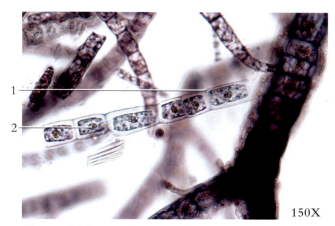

Figure 4.54 Magnified view of thallus of *Ectocarpus* sp. showing cellular details.

1. Cell wall
2. Nucleus

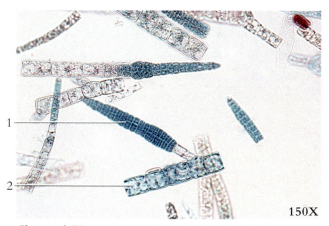

Figure 4.55 *Ectocarpus* sp. showing pleurolocular sporangia.
1. Pleurolocular sporangium
2. Filament of cells

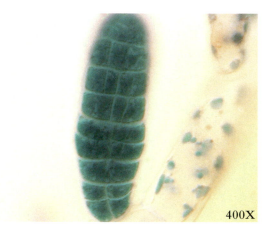

Figure 4.56 Magnified view of *Ectocarpus* sp. showing pleurolocular sporangium.

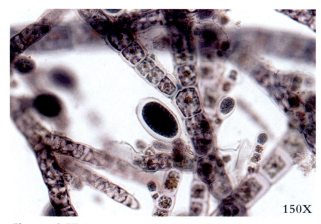

Figure 4.57 *Ectocarpus* sp. showing unilocular sporangia.

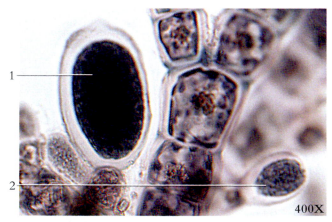

Figure 4.58 Magnified view of *Ectocarpus* sp. showing unilocular sporangia.

1. Mature unilocular sporangium
2. Immature unilocular sporangium

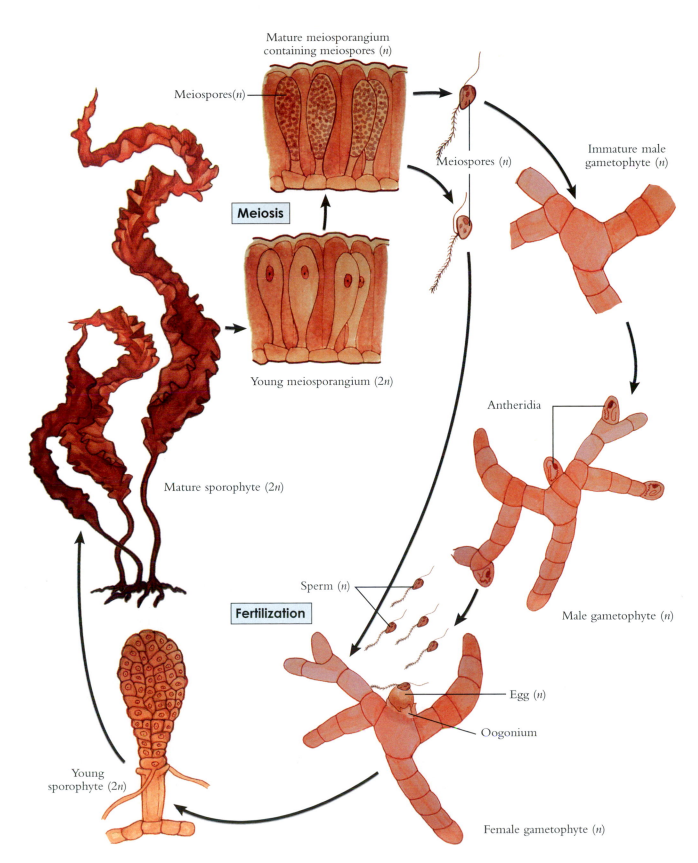

Mature meiosporangium
containing meiospores (n)

Meiospores(n)

Meiosis

Young meiosporangium (2n)

Meiospores (n)

Immature male
gametophyte (n)

Antheridia

Mature sporophyte (2n)

Male gametophyte (n)

Sperm (n)

Fertilization

Egg (n)

Oogonium

Young
sporophyte (2n)

Female gametophyte (n)

Figure 4.59 Life cycle of *Laminaria*, a common kelp.

Figure 4.60 Rocky coast of southern Alaska showing dense growths of the brown alga, *Fucus* sp.

Figure 4.61 "Sea palm," *Postelsia palmaeformis*, a common brown alga found on the western coast of North America.

Macrocystis sp.

Macrocystis sp.

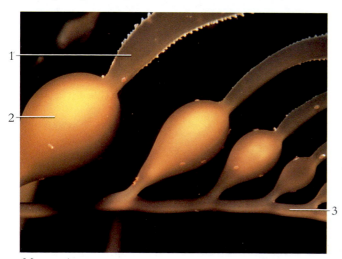

Macrocystis sp.

Egregia sp.

Figure 4.62 Examples of brown algae, Phaeophyta. These large species are commonly known as kelps.

1. Blade   3. Stipe
2. Float (air-filled bladder)

Figure 4.63 Kelp, *Laminaria* sp., is one of the common "sea weeds" found along many rocky coasts.

Figure 4.64 Tidal pool with green, brown and red algae.

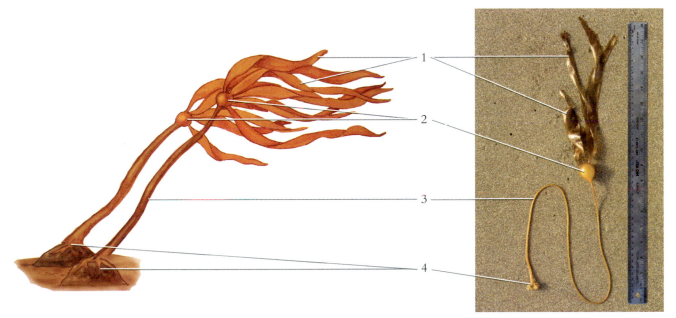

Figure 4.65 Brown alga, *Nereocystis sp.,* has a long stipe and photosynthetic laminae attached to a large float. The holdfast anchors the alga to the ocean floor. This alga and others can grow to lengths of several meters.

1. Lamina 2. Floats (air-filled bladders) 3. Stipe 4. Holdfasts

Figure 4.66 *Sargassum* sp., a brown alga common in the Sargasso sea.

1. Floats 2. Blade 3. Stipe

Figure 4.67 Mixture of kelps washed onto shore to form "windrows" of *Phaeophyta* sp.

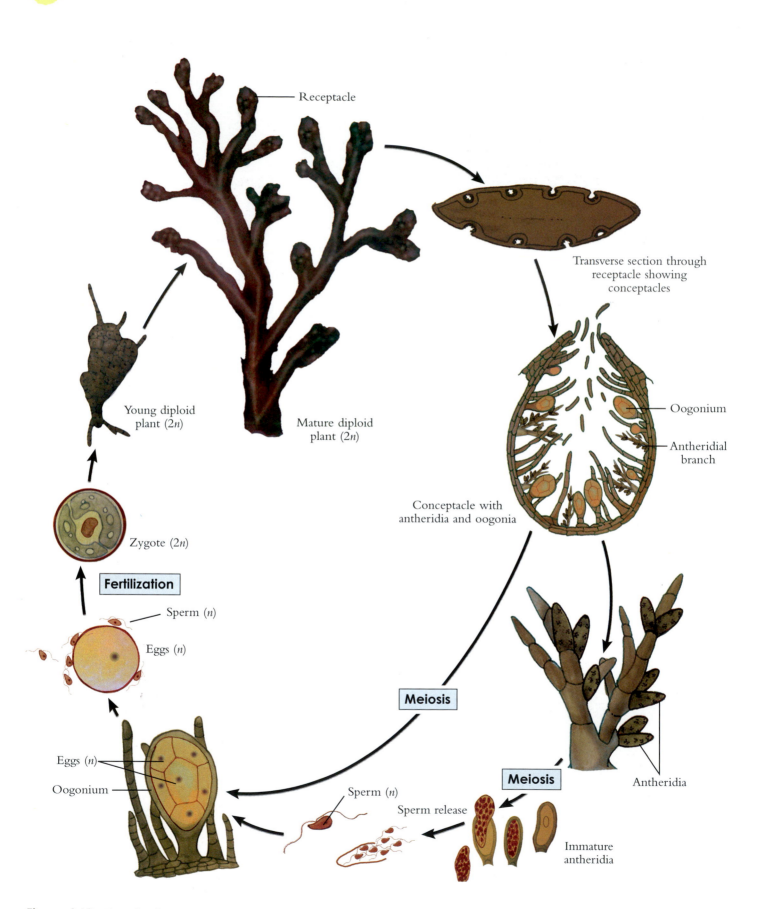

Receptacle

Transverse section through receptacle showing conceptacles

Young diploid plant (2*n*)

Mature diploid plant (2*n*)

Oogonium

Antheridial branch

Conceptacle with antheridia and oogonia

Zygote (2*n*)

Fertilization

Sperm (*n*)

Eggs (*n*)

Meiosis

Eggs (*n*)

Oogonium

Meiosis

Sperm (*n*)

Sperm release

Antheridia

Immature antheridia

Figure 4.68 Life cycle of *Fucus* sp., a common brown alga.

Figure 4.69 (a) *Fucus* sp., a brown alga, commonly called rockweed. (b) An enlargement of a blade supporting the receptacles.

1. Blade
2. Receptacle
3. Conceptacles (spots) are chambers imbedded in the receptacles
4. Blade

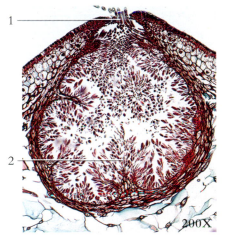

Figure 4.70 Conceptacle of *Fucus* sp.
1. Sterile paraphyses
2. Antheridial branches

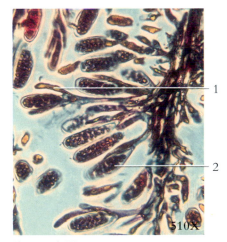

Figure 4.71 Close-up of antheridial branch of *Fucus* sp.
1. Antheridial branch
2. Antheridium

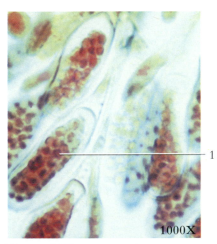

Figure 4.72 Close-up of antheridium of *Fucus* sp.
1. Sperm within antheridium

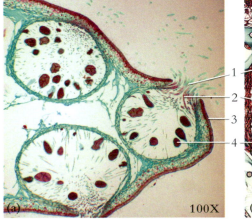

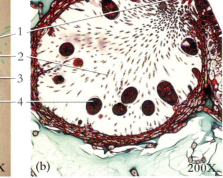

Figure 4.73 Section through a *Fucus* sp. receptacle. (a) Low magnification showing three conceptacles and (b) higher magnification of a single conceptacle with oogonia.
1. Ostiole
2. Paraphyses (sterile hairs)
3. Surface of receptacle
4. Oogonium

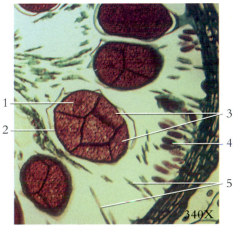

Figure 4.74 *Fucus* sp., close-up of bisexual conceptacle.
1. Nucleus of egg
2. Oogonium
3. Eggs
4. Antheridium
5. Paraphyses

Phylum Rhodophyta - red algae

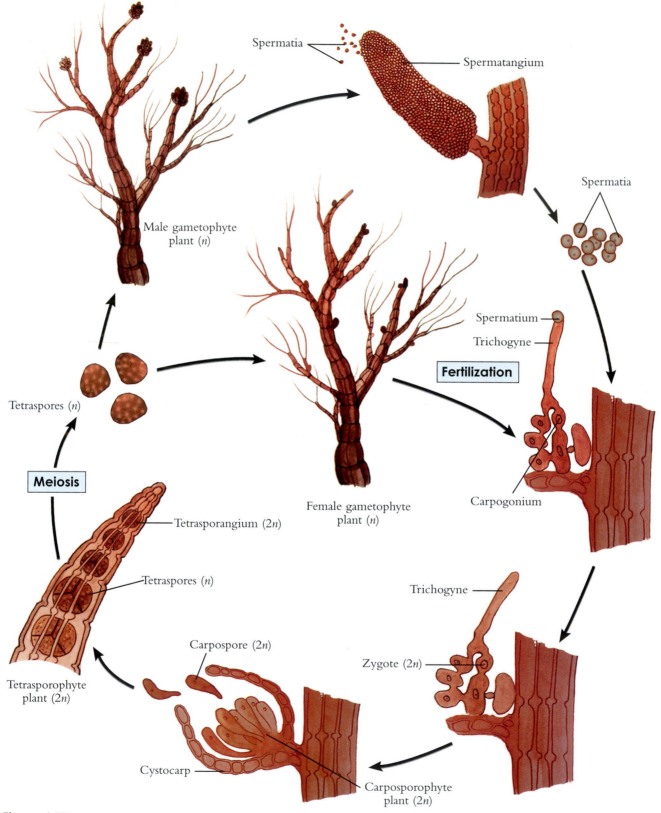

Figure 4.75 Life cycle of the red alga, *Polysiphonia* sp.

Figure 4.76 Intertidal zone showing a colony of a filamentous red alga, *Bangia* sp.

Figure 4.77 Mature plant of the red alga, *Rhodymenia* sp.

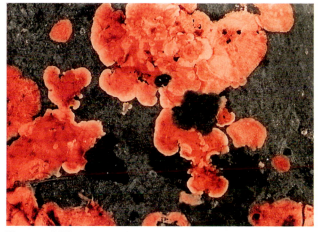

Figure 4.78 Small encrusting colonies of a species of red alga on a stone. The colonies shown are bright red and are only a few millimeters in size.

250X

Figure 4.79 *Batrachospermum* sp., a common fresh-water red alga.

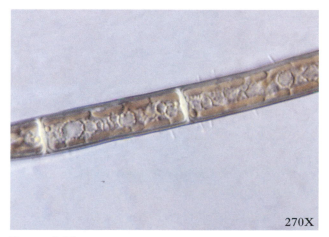

270X

Figure 4.80 *Audouinella* sp. is a fresh-water member of Rhodophyta. This organism was collected from a cold-water spring.

Figure 4.81 Mature plant of the common red alga, *Polysiphonia* sp.

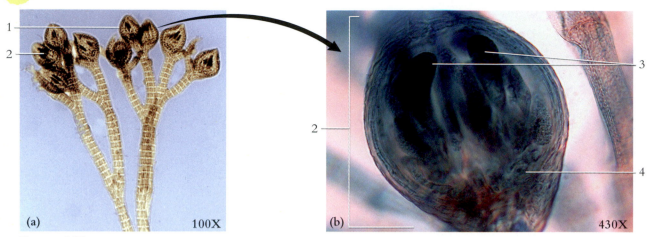

Figure 4.82 Red alga, *Polysiphonia* sp., has alternation of three generations. (a) Female gametophyte with attached carposporophyte generation. (b) A closeup of cystocarp containing a carposporophyte.

  1. Pericarp   2. Cystocarp   3. Carpospores   4. Carposporophyte

Figure 4.83 *Polysiphonia* sp., showing the release of carpospores.

  1. Carpospores (2*n*)   2. Ruptured cystocarp

Figure 4.84 Tetrasporophyte generation of *Polysiphonia* sp. showing tetraspores (meiospores).

  1. Tetraspores   2. Cells of tetrasporophyte plant

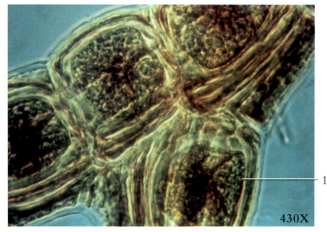

Figure 4.85 Close-up of tetrasporophyte plant of *Polysiphonia* sp.

  1. Tetraspore (meiospore)

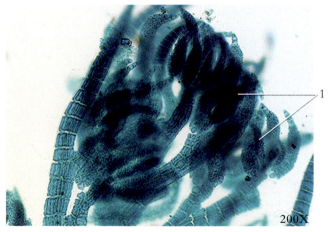

Figure 4.86 Male gametophyte plant of *Polysiphonia* sp., showing spermatangia (green stain).

  1. Spermatangia with spermatia

Phylum Myxomycota - plasmodial slime molds

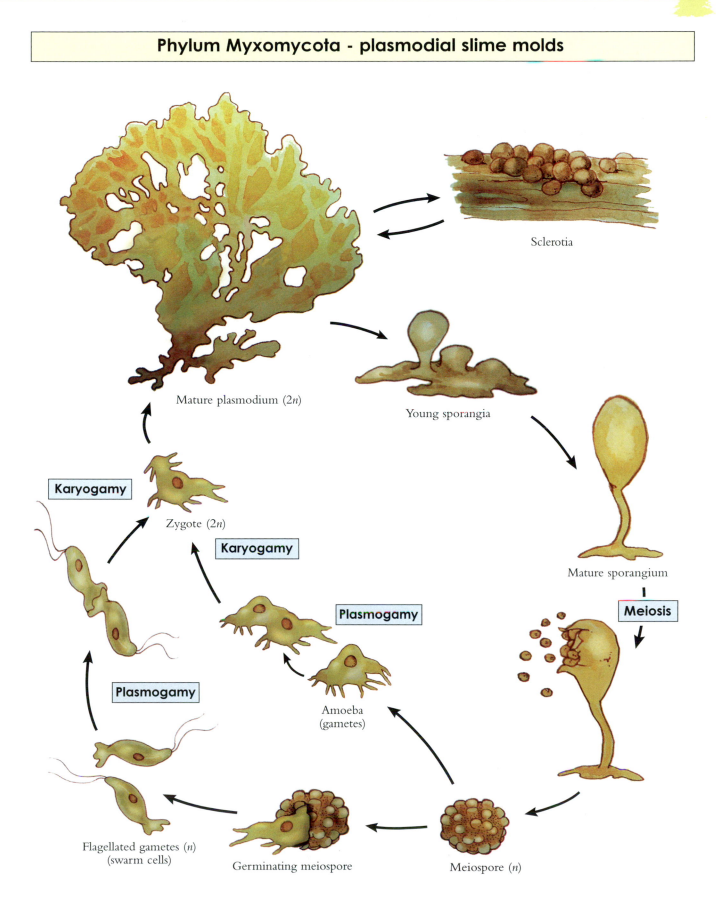

Mature plasmodium (2*n*)

Sclerotia

Young sporangia

Mature sporangium

Meiosis

Karyogamy

Zygote (2*n*)

Karyogamy

Plasmogamy

Amoeba
(gametes)

Plasmogamy

Meiospore (*n*)

Flagellated gametes (*n*)
(swarm cells)

Germinating meiospore

Figure 4.87 Life cycle of a plasmodial slime mold.

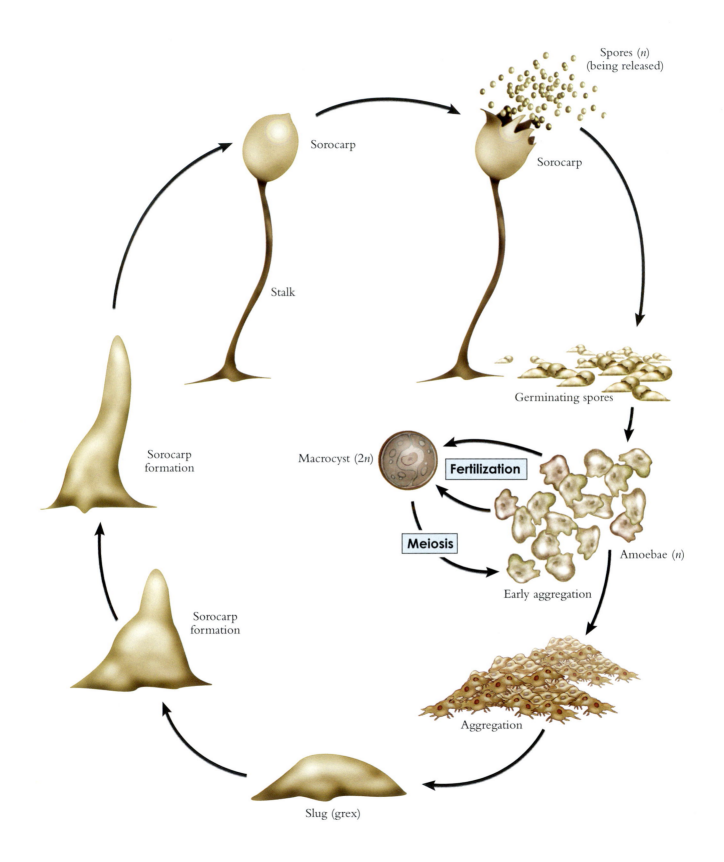

Figure 4.88 Life cycle of the cellular slime mold *Dictyostelium* sp.

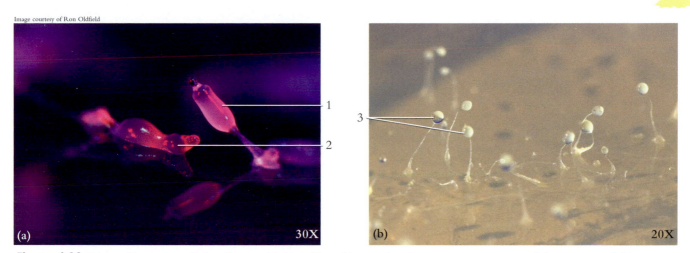

Image courtesy of Ron Oldfield

(a)   30X

(b)   20X

Figure 4.89 *Dictyostelium* sp., a cellular slime mold. (a) A filtered image showing a young sorocarp and slug stages and (b) true color showing mature sorocarps.

   1. Developing sorocarp   2. Slug (grex)   3. Mature sorocarps

Figure 4.90 Aethalium (sporangium) of the plasmodial slime mold *Fuligo septica*, often called the dog vomit fungus (scale in mm).

Figure 4.91 Plasmodial slime mold sporangia on underside of burned log in Yellowstone National Park.

Figure 4.92 Plasmodial slime mold *Physarum cinerea* showing (a) developing sporania on blue grass leaves and (b) mature sporangia with black spores and lime crust at same location.

Figure 4.93 Sporangia of the slime mold *Comatricha typhoides*.

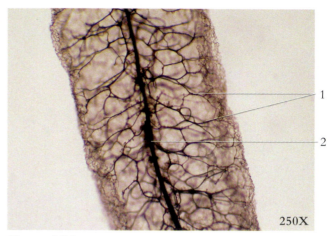

Figure 4.94 Longitudinal section through the sporangium of *Stemonitis* sp.
1. Cellular filaments (capillitia) 2. Columella

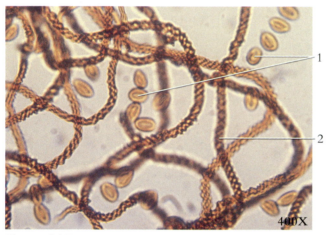

Figure 4.95 Close-up through the sporangium of *Stemonitis* sp.
1. Spores 2. Capillitia

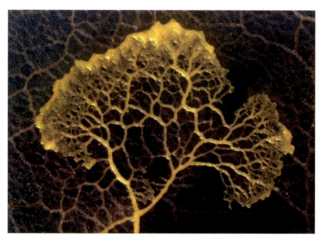

Figure 4.96 *Physarum* sp. plasmodium.

Figure 4.97 Slime mold sporangia vary considerably in size and shape. One species of *Lycogala* is shown here.

Figure 4.98 Slime mold specimen from a high mountain locality.

Phylum Oomycota - water molds, white rusts, and downy mildews

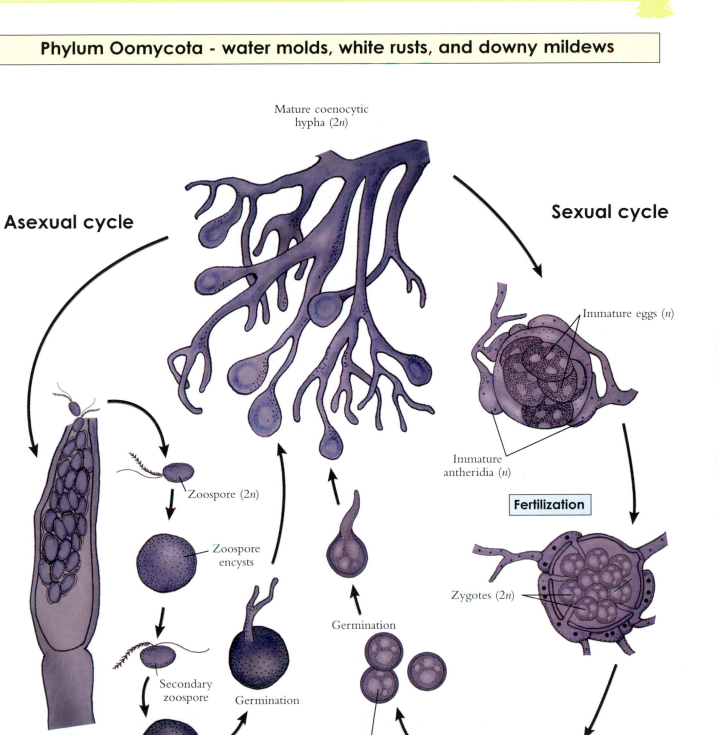

Mature coenocytic
hypha (2n)

Asexual cycle

Sexual cycle

Immature eggs (n)

Immature
antheridia (n)

Zoospore (2n)

Zoospore
encysts

Fertilization

Zygotes (2n)

Germination

Secondary
zoospore

Germination

Zoosporangium

Oospores (2n)

Secondary
zoospore encysts

Figure 4.99 Life cycle of the water mold *Saprolegnia* sp.

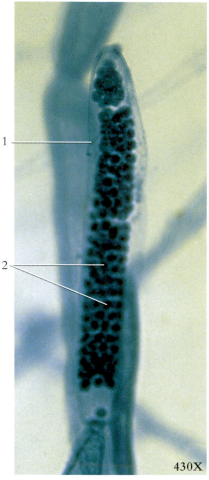

430X

Figure 4.100 Zoosporangium of the
water mold *Saprolegnia* sp.
1. Zoosporangium
2. Zoospores

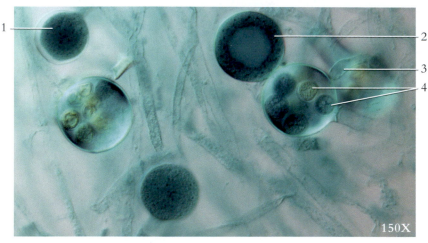

150X

Figure 4.101 Oogonia of the water mold *Saprolegnia* sp.
1. Young oogonium   3. Young antheridium
2. Developing oogonium   4. Eggs

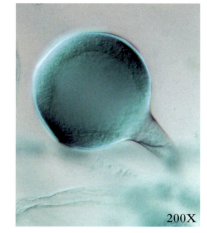

200X

Figure 4.102 Water mold,
Saprolegnia sp., showing a young
oogonium before eggs have
been formed.

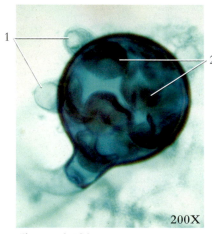

200X

Figure 4.103 Oogonia of the
water mold *Saprolegnia* sp.
1. Empty antheridia
2. Zygotes

Figure 4.104 Skin of this brown trout has been infected by the common water mold, *Saprolegnia* sp.

Chapter 5

Fungi, Yeast, Molds, Mushrooms, Rusts, and Lichens

About 250,000 species of fungi are currently extant on Earth. All fungi are heterotrophs; they absorb nutrients through their cell walls and cell membranes. The kingdom Fungi includes the conjugation fungi, yeasts, mushrooms, toadstools, rusts, and lichens. Most are saprobes, absorbing nutrients from dead organic material. Some are parasitic, absorbing nutrients from living hosts. Fungi decompose organic material, helping to recycle nutrients essential for plant growth.

Except for the unicellular yeasts, fungi consist of elongated filaments called *hyphae*. Hyphae begin as cellular extensions of spores that branch as they grow to form a network of hyphae called a *mycelium*. Even the body of a mushroom consists of a mass of tightly packed hyphae attached to an underground mycelium. Fungi are nonmotile and reproduce by means of spores, produced sexually or asexually.

Many species of fungi are commercially important. Some are used as food, such as mushrooms; or in the production of foods, such as bread, cheese, beer, and wine. Other species are important in medicine, for example, in the production of the antibiotic penicillin. Many other species of fungi are of medical and economic concern because they cause plant and animal diseases and destroy crops and stored goods.

Table 5.1 Some Representatives of Fungi

Phyla and Representative Kinds	Characteristics
Zygomycota — bread molds, fly fungi	Hyphae lack cross walls along filaments; sexual reproduction by conjugation
Ascomycota — yeasts, molds, morels, and truffles	Septate hyphae; reproductive structure contains ascospores within asci in fruiting bodies known as ascocarps (ascoma); asexual reproduction by budding or conidia
Basidiomycota — mushrooms, toadstools, rusts, and smuts	Septate hyphae; 4 meiospores produced externally on cells called basidia contained in basidiocarp (basidioma)
Lichens — not a phylum, but a symbiotic association of an alga and a fungus	Algal component (usually a green alga) provides food from photosynthesis; fungal component (usually an ascomycete) may provide anchorage, water retention, and/or nutrient absorbance

Figure 5.1 Examples of common fungi. (a) A cup fungus, *Peziza* sp., a common ascomycete, and (b) oyster mushrooms, *Pleurotus* sp.

Phylum Zygomycota - conjugation fungi

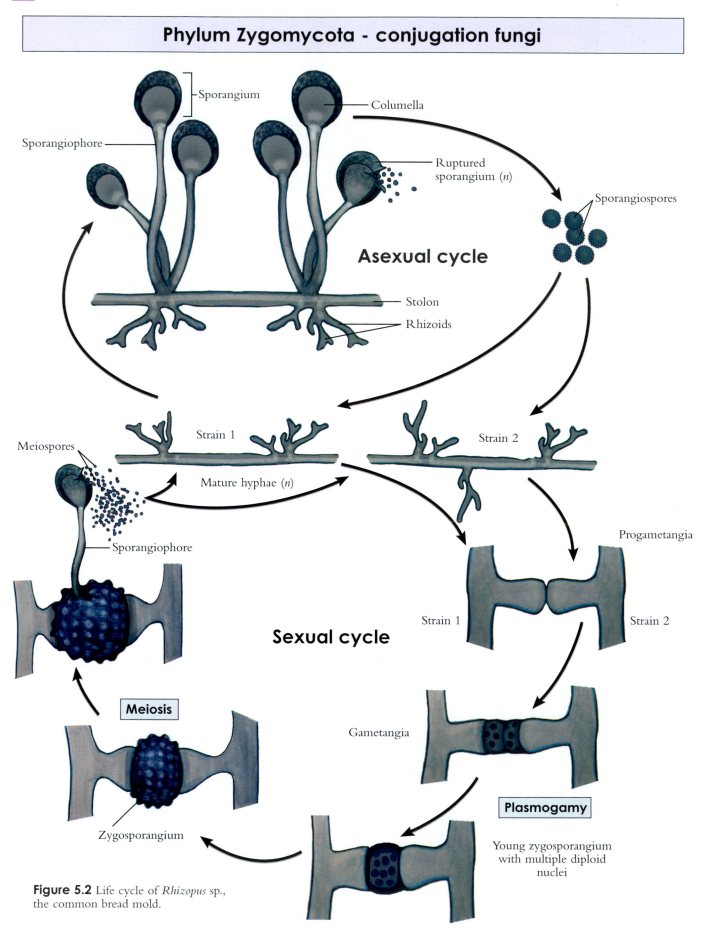

Sporangium

Columella

Sporangiophore

Ruptured sporangium (n)

Sporangiospores

Asexual cycle

Stolon

Rhizoids

Meiospores

Strain 1

Strain 2

Mature hyphae (n)

Progametangia

Sporangiophore

Strain 1

Strain 2

Sexual cycle

Meiosis

Gametangia

Plasmogamy

Zygosporangium

Young zygosporangium with multiple diploid nuclei

Figure 5.2 Life cycle of *Rhizopus* sp., the common bread mold.

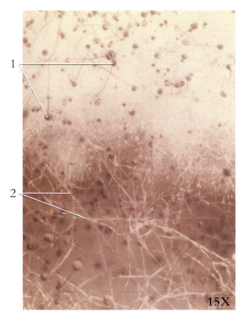

Figure 5.3 *Rhizopus* sp. growing on a slice of bread.
  1. Sporangia
  2. Hyphae (stolon)

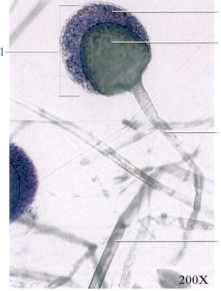

Figure 5.4 Whole mount of the bread mold, *Rhizopus* sp.
  1. Sporangium
  2. Spores
  3. Columella
  4. Sporangiophore
  5. Hypha

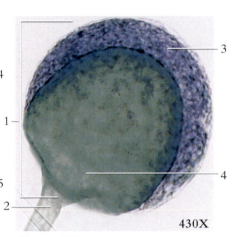

Figure 5.5 Mature sporangium in the asexual reproductive cycle of the bread mold, *Rhizopus* sp.
  1. Sporangium
  2. Sporangiophore
  3. Spores
  4. Columella

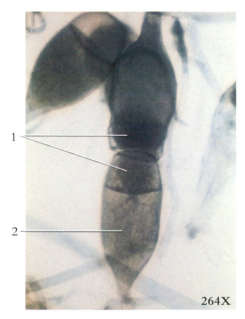

Figure 5.6 Young gametangia of *Rhizopus* sp. contacting prior to plasmogamy.
  1. Immature gametangia
  2. Suspensor cell

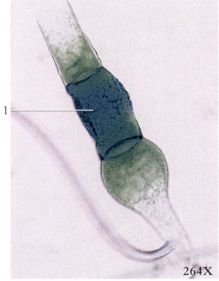

Figure 5.7 Immature *Rhizopus* sp. zygospore following plasmogamy.
  1. Immature zygosporangium

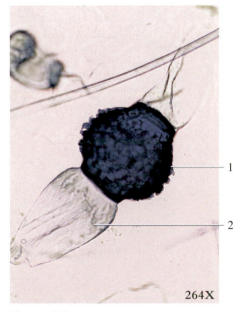

Figure 5.8 Mature *Rhizopus* sp. zygospore.
  1. Zygosporangium
  2. Suspensor cell

Phylum Ascomycota -yeasts, molds, morels, and truffles

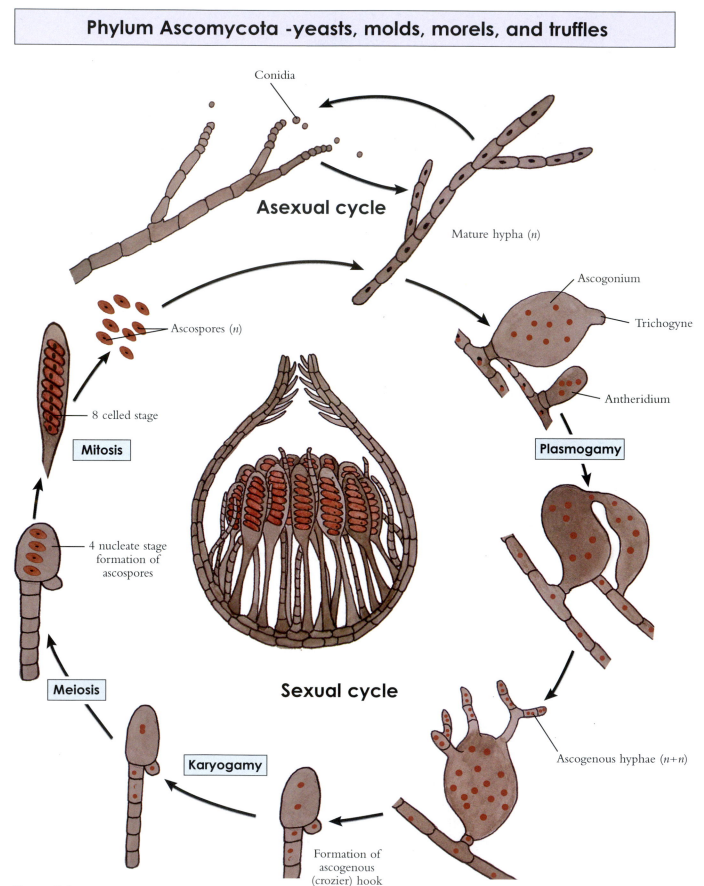

Conidia

Asexual cycle

Mature hypha (*n*)

Ascogonium

Trichogyne

Ascospores (*n*)

Antheridium

8 celled stage

Mitosis

Plasmogamy

4 nucleate stage
formation of
ascospores

Sexual cycle

Meiosis

Ascogenous hyphae (*n*+*n*)

Karyogamy

Formation of
ascogenous
(crozier) hook

Figure 5.9 Life cycle of an ascomycete.

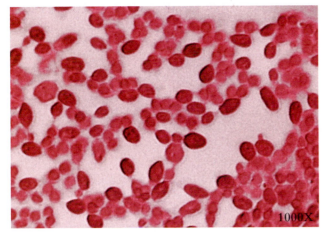

Figure 5.10 Baker's yeast, *Saccharomyces cerevisiae.* The ascospores of this unicellular ascomycete are characteristically spheroidal or ellipsoidal in shape.

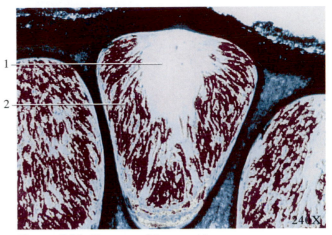

Figure 5.11 Close-up of the parasitic ascomycete, *Hypoxylon* sp., showing imbedded perithecia.
1. Perithecium   2. Hymenium

Figure 5.12 Parasitic ascomycete, *Dibotryon morbosum*, on a branch of a choke cherry, *Prunus virginiana.*
1. Fungus   2. Choke cherry stem

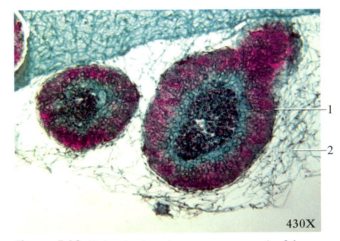

Figure 5.13 Cleistothecium (ascocarp or ascoma) of the ascomycete *Penicillium* sp.
1. Cleistothecium
2. Hyphae

Peziza repanda   *Scutellinia scutellata*   *Morchella* sp.   *Helvella* sp.

Figure 5.14 Fruiting bodies (ascocarps or ascoma) of common ascomycetes. *Peziza repanda* is a common woodland cup fungus. *Scutellinia scutellata* is commonly called the eyelash cup fungus. *Morchella esculenta* is a common edible morel. *Helvella* is sometimes known as a saddle fungus since the fruiting body is thought by some to resemble a saddle.

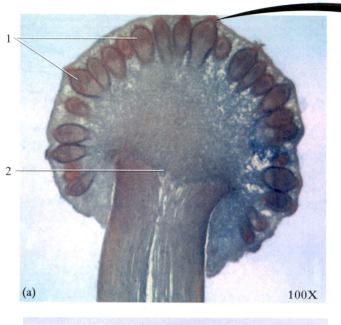

(a)   100X

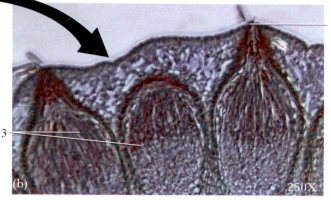

(b)   250X

Figure 5.15 Ascomycete, *Claviceps purpurea*.
(a) Longitudinal section through stroma showing ascocarps (ascoma). (b)Enlargement of three perithecia. This fungus causes serious plant diseases and is toxic to humans.

1. Perithecia
2. Multiple perithecia within stroma
3. Perithecia containing asci
4. Ostiole

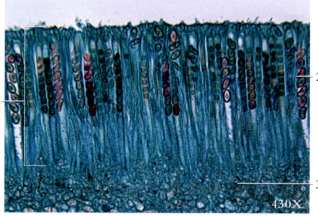

430X

Figure 5.16 Section through the hymenial layer of the apothecium of *Peziza* sp., showing asci with ascospores.

1. Hymenial layer
2. Ascus with ascospores
3. Ascocarp (ascoma) mycelium

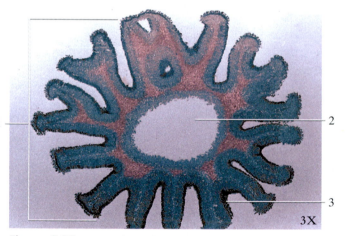

3X

Figure 5.17 Section through an ascocarp (ascoma) of the morel, *Morchella* sp.. True morels are prized for their excellent flavor.

1. Convoluted fruiting body
2. Hollow "stalk"
3. Hymenium

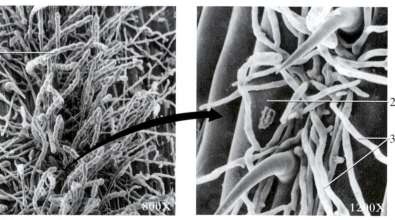

800X   1200X

Figure 5.18 Scanning electron micrographs of the powdery mildew, *Erysiphe graminis*, on the surface of wheat. As the mycelium develops, it produces spores (conidia) that give a powdery appearance to the wheat.

1. Conidia
2. Wheat host
3. Hyphae of the fungus

1800X

Photograph courtesy of: James V. Allen

Figure 5.19 Scanning electron micrograph of a germinating spore (conidium) of the powdery mildew, *Erysiphe graminis*. The spore develops into a mycelium that penetrates the epidermis and then spreads over the host plant producing a powdery appearance.

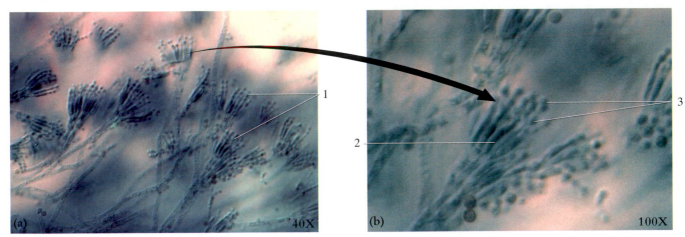

Figure 5.20 Fungus *Penicillium* sp. causes economic damage as a mold but is also the source of important antibiotics. (a) A colony of *Penicillium* sp., and (b) a close-up of a conidiophore with chains of asexual spores (conidia) at the end.

  1. Conidia   2. Conidiophore   3. Conidia

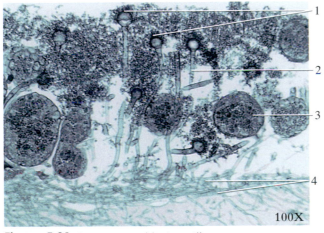

Figure 5.21 Common mold, *Aspergillus* sp.
  1. Conidia (spores)   3. Cleistothecium
  2. Conidiophore   4. Hyphae

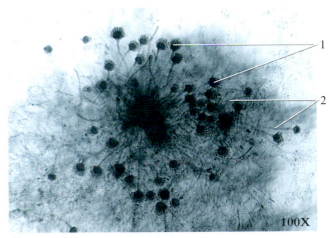

Figure 5.22 Common mold, *Aspergillus* sp.
  1. Conidia   2. Conidiophores

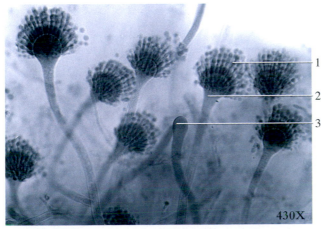

Figure 5.23 Close-up of sporangia of the mold, *Aspergillus* sp. The conidia, or spores, of this genus are produced in a characteristic radiate pattern.
  1. Conidia (spores)   3. Developing conidiophore
  2. Conidiophore

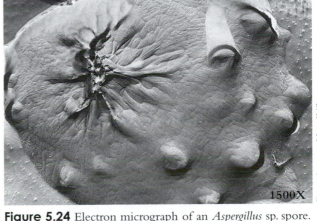

Figure 5.24 Electron micrograph of an *Aspergillus* sp. spore. Note the rodlet pattern on the spore wall.

Photograph courtesy of: James V. Allen

Phylum Basidiomycota - mushrooms, toadstools, rusts, and smuts

Pleurotus sp.

Hericium sp.

Coriolus sp.

Astreus sp.

Coprinus sp.

Amanita sp.

Chantarella sp.

Amanita sp.

Nidularia sp.

Boletus sp.

Figure 5.25 Representative basidiocarps (basidiomas or fruiting bodies) of basidiomycetes.

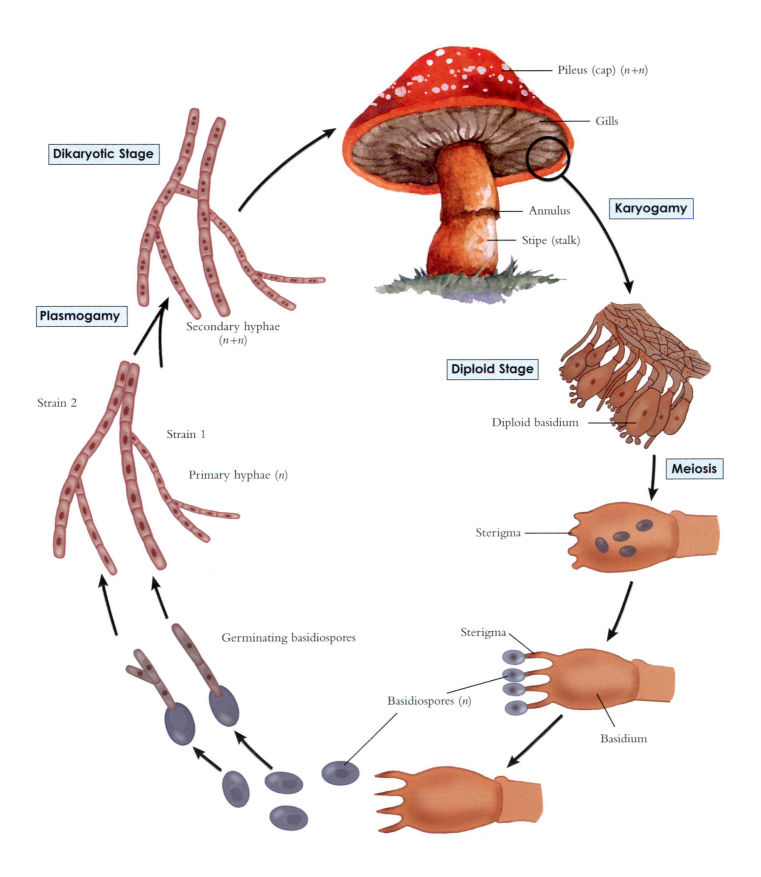

Dikaryotic Stage

Pileus (cap) (*n+n*)

Gills

Karyogamy

Annulus

Stipe (stalk)

Plasmogamy

Secondary hyphae
(*n+n*)

Diploid Stage

Strain 2

Diploid basidium

Strain 1

Meiosis

Primary hyphae (*n*)

Sterigma

Germinating basidiospores

Sterigma

Basidiospores (*n*)

Basidium

Figure 5.26 Life cycle of a "typical" basidiomycete (mushroom).

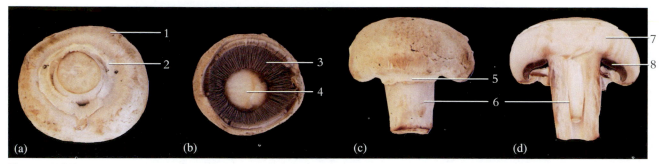

(a)   (b)   (c)   (d)

Figure 5.27 Structure of a mushroom. (a) An inferior view with the annulus intact, (b) an inferior view with the veil removed to show the gills, (c) a lateral view, and (d) a longitudinal section.

1. Pileus (cap)   4. Stipe (stalk)   7. Pileus (cap)
2. Veil   5. Annulus   8. Gills
3. Gills   6. Stipe (stalk)

Figure 5.28 Basidiomycete puffballs growing on a decaying log.

Figure 5.29 Herbarium specimen of the wood fungus, *Stropharia semiglobata*. Growing on decaying wood and other organic matter, basidiomycetes are important decomposers in forest communities.

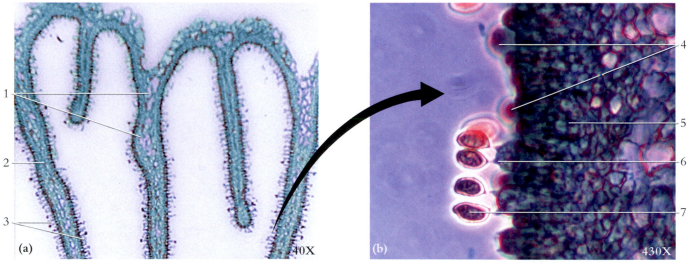

(a)   40X   (b)   430X

Figure 5.30 Gills of the mushroom *Coprinus* sp. (a) A close-up of several gills, and (b) a close-up of a single gill.

1. Hyphae comprising the gills   4. Immature basidia   7. Basidiospore
2. Gill   5. Gill (comprised of hyphae)
3. Basidiospores   6. Sterigma

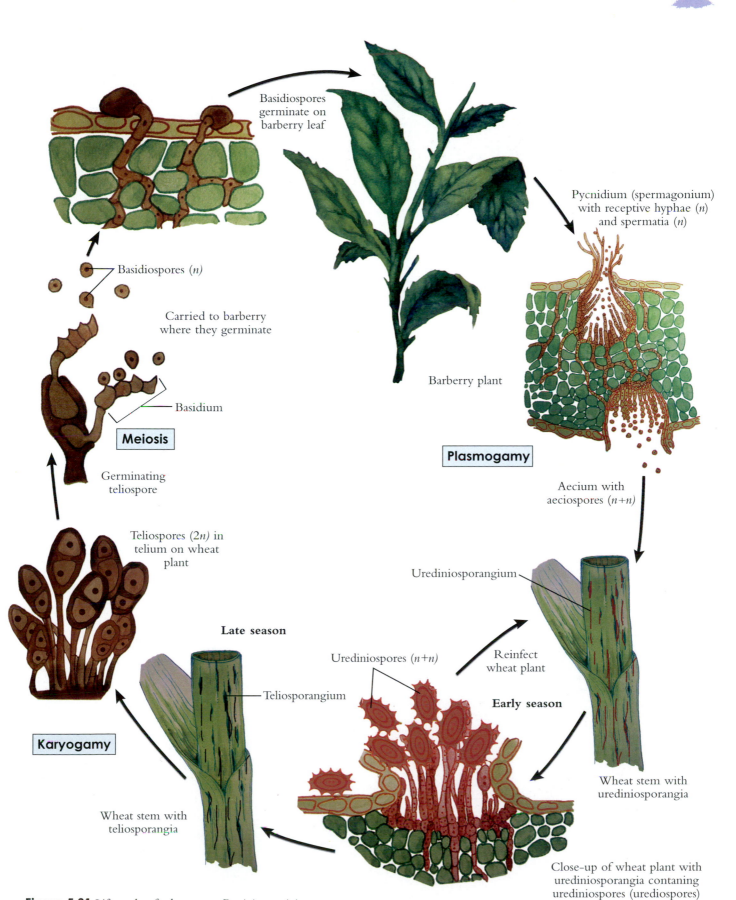

Figure 5.31 Life cycle of wheat rust, *Puccinia graminis.*

Basidiospores germinate on barberry leaf

Basidiospores (*n*)

Carried to barberry where they germinate

Basidium

Meiosis

Germinating teliospore

Teliospores (*2n*) in telium on wheat plant

Karyogamy

Wheat stem with teliosporangia

Teliosporangium

Late season

Pycnidium (spermagonium) with receptive hyphae (*n*) and spermatia (*n*)

Barberry plant

Plasmogamy

Aecium with aeciospores (*n+n*)

Urediniosporangium

Reinfect wheat plant

Urediniospores (*n+n*)

Early season

Wheat stem with urediniosporangia

Close-up of wheat plant with urediniosporangia contaning urediniospores (urediospores)

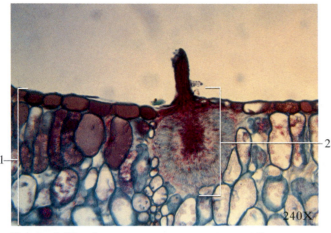

Figure 5.32 Wheat rust, *Puccinia graminis,* pycnidium on barberry leaf.

1. Barberry leaf   2. Pycnidium (spermagonium)

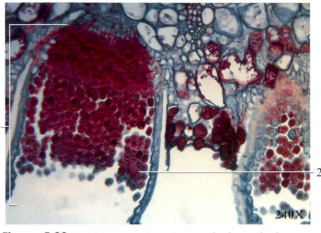

Figure 5.33 *Puccinia graminis,* aecium on barberry leaf.

1. Aecium   2. Aeciospores

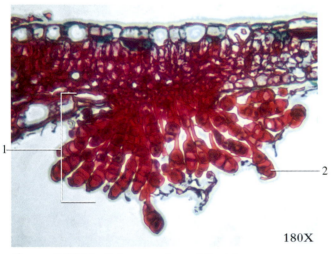

Figure 5.34 Urediniosporangium of *Puccinia* sp. on wheat leaf.

1. Urediniosporangium   2. Urediniospore

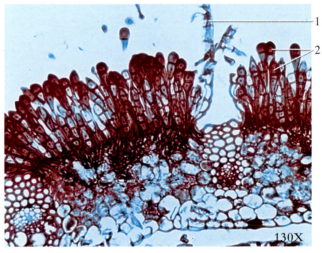

Figure 5.35 Close-up of a wheat leaf sheath showing telia of wheat rust, *Puccinia graminis.*

1. Epidermis of leaf   2. Teliospores

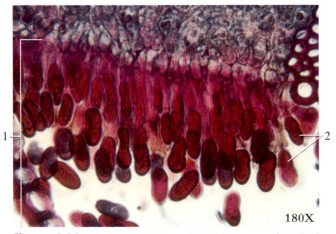

Figure 5.36 Close-up of telium of *Puccinia* sp. on wheat leaf.

1. Telium
2. Teliospores

Figure 5.37 May apple rust, *Puccinia podophylli,* showing aecia on the lower surface of May apple leaves.

1. Clusters of aecia

Figure 5.38 Black stem wheat rust, *Puccinia graminis,* on the lower surface of barberry leaves.
1. Clusters of aecia

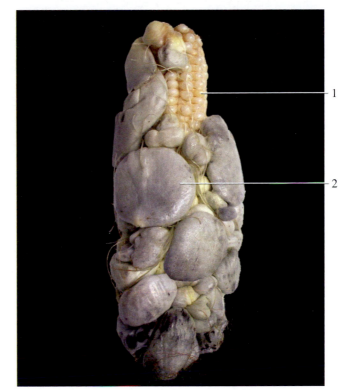

Figure 5.39 Corn ear, *Zea mays*, infected by the smut *Ustilago maydis,* which is destroying the fruit (ear).
1. Corn ear
2. Fungus

Figure 5.40 Smut-infected brome grass. The grains have been destroyed by the smut fungus.

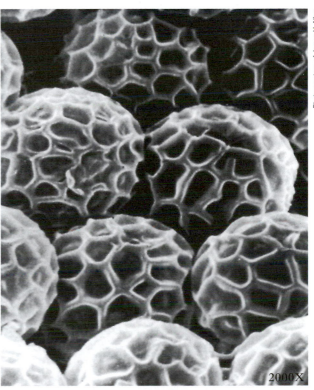

Photo courtesy of James V. Allen

2000 X

Figure 5.41 Scanning electron micrograph of teliospores of a wheat smut fungus.

Lichens (symbiotic associations of fungi and algae)

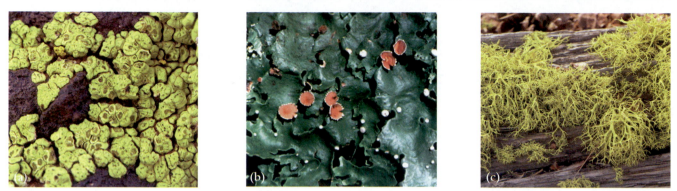

Figure 5.42 Lichens are often characterized informally on the basis of their form. (a) Crustose lichen, (b) foliose lichen, and (c) fruticose lichen.

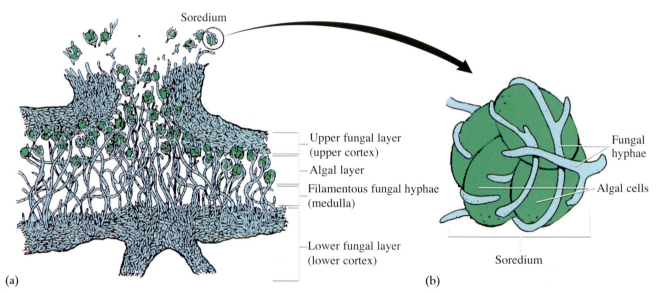

Figure 5.43 Many lichens reproduce by producing soredia, which are small bodies containing both algal and fungal cells. (a) Lichen thallus, and (b) soredium.

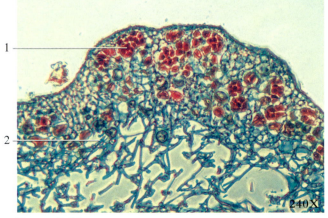

Figure 5.44 Transverse section through a lichen thallus.
1. Algal cells   2. Fungal hyphae

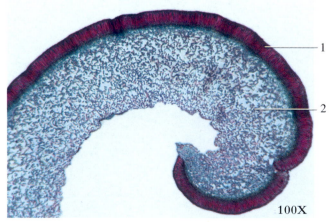

Figure 5.45 Ascomycete lichen thallus demonstrating a surface layer of asci.
1. Asci   2. Loose fungal filaments

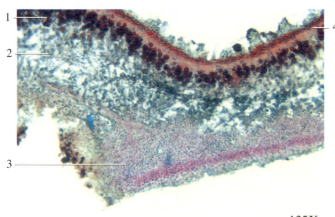

125X

Figure 5.46 Transverse section through a lichen thallus.
1. Algal cells   3. Lower cortex
2. Medulla   4. Fungal layer (upper cortex)

Figure 5.47 Foliose lichen *Xanthoria* sp. growing on the bark of a tree.
1. Lichen   2. Bark

Figure 5.48 Crustose lichen *Lecanora* sp. growing on sandstone in an arid Southern Utah environment.

Figure 5.49 Foliose lichen *Hypogymnia* sp. growing on a pine branch in the Pacific Northwest.

Figure 5.50 Fruticose lichen *Alectoria* sp. growing in Northern California.

Figure 5.51 Foliose and fruticose lichens in the Pacific Northwest.
1. Foliose lichen *Hypogymnia* sp.
2. Foliose lichen *Evernia* sp.
3. Fruticose lichen *Usnea* sp.

Chapter 6
Liverworts, Hornworts, and Mosses

The bryophytes include the liverworts, hornworts, and mosses. Many bryologists include them in one phylum with three classes: Hepaticae, Anthocerotae, and Musci (mosses only). Others divide this group into three phyla: Hepatophyta, Anthocerophyta, and Bryophyta. These plants inhabit damp, shady environments and are found worldwide. Though simple plants, bryophytes are old and successful. Late Silurian to Devonian fossil bryophytes have been found, some 375 to 400 million years old. Currently, some 16,000 extant species are known. Because many bryophytes are sensitive to sulfur dioxide, they cannot live in air polluted environments.

Although most bryophytes are only 1 to 2 cm in height, a few species may grow to 15 to 20 cm. The adult, free-living plant is haploid, produces gametes, and is referred to as a gametophyte. Gametophytes of many bryophytes have a thick, waxy *cuticle* that protects the plant from desiccation. Lacking roots, gametophytes are attached to the substrate by elongated single cells or multicellular filaments known as *rhizoids*. Rhizoids of gametophytes often extend horizontally intertwining with rhizoids of other gametophytes, forming loose colonies of individual plants that are effective in holding moisture.

In some liverworts and hornworts the gametophyte is dorsoventrally flattened, while in some liverworts it is upright and "leafy." With the possible exception of certain mosses, bryophytes lack vascular tissue. Transport within the stem is through diffusion, capillary action, and cytoplasmic streaming. Lacking true leaves, some bryophytes have leaf-like extensions that collect moisture and assist in reproduction. *Stomata* (for gas exchange) surrounded by two guard cells are present on the sporophytes of hornworts and mosses. Unlike stomata of typical flowering plants, however, the stomata of some bryophytes are openings surrounded by a single, doughnut-shaped binucleated guard cell.

Like all true plants, bryophytes have *alternation of generations*. This means their reproductive cycle has a haploid (n) phase in which the gametophyte produces gametes. After fusion in pairs, the gametes form a *zygote*. The zygote germinates, producing a diploid ($2n$) that matures to become the sporophyte. Through meiotic division, the spores produced from the sporophytes complete the cycle by giving rise to new gametophytes.

Most bryophytes have separate male and female gametophytes. The male gametophytes have antheridia which produce flagellated sperm cells. The female gametophytes have archegonia in which eggs are produced.

Water is essential for fertilization because sperm produced by the antheridia swim to the archegonia. In some species of bryophytes, raindrops may disperse the sperm and insects may play a limited role in dispersal. The diploid zygote develops into an embryonic sporophyte within the protective jacket of the archegonium.

During sporophyte development a *stalk,* or *seta,* forms to free the plant from the archegonium. A spore-producing capsule, or *sporangium*, forms at the tip of the stalk. The spores, produced through meiosis, are haploid cells that disperse when the sporangium bursts. As the spores germinate, a threadlike protonema is produced that gives rise to a haploid gametophyte, completing the life cycle.

Figure 6.1 *Marchantia*, a common liverwort.

Table 6.1 The Three Phyla of Bryophytes

Phyla	Characteristics
Hepatophyta — liverworts	Flat or leafy gametophytes; single-celled rhizoids; simple sporophytes and elaters present; stomata and columella absent
Anthocerophyta — hornworts	Flat, lobed gametophytes; more complex sporophytes with stomata; pseudoelaters and columella present
Bryophyta — mosses	Leafy gametophytes, multicellular rhizoids; sporophytes with stomata, columella, peristome teeth and/or operculum present

Phylum Hepatophyta - liverworts

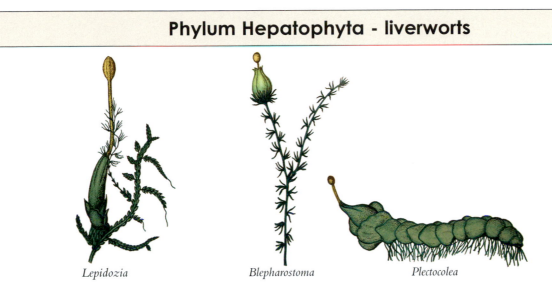

Lepidozia   *Blepharostoma*   *Plectocolea*

Figure 6.2 Illustration of three genera of leafy liverworts, showing the gametophyte with an attached sporophyte. The perianth contains the archegonia and the lower portion of the developing sporophyte (yellowish).

Calypogeia sp.

Conocephalum sp.

Bazzania sp.

Porella sp.

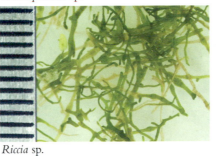

Riccia sp.

Scapania sp.

Figure 6.3 Examples of liverworts (scale in mm).

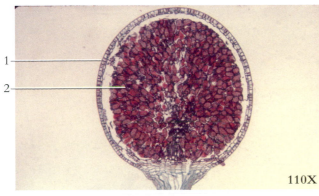

110X

Figure 6.4 Sporophyte (capsule) of the leafy liverwort, *Pellia* sp..

1. Capsule   2. Sporogenous tissue

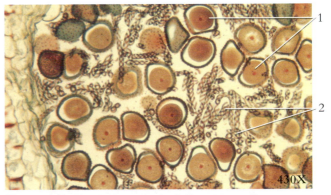

430X

Figure 6.5 Capsule from the leafy liverwort, *Pellia* sp., in longitudinal view.

1. Spores   2. Elaters

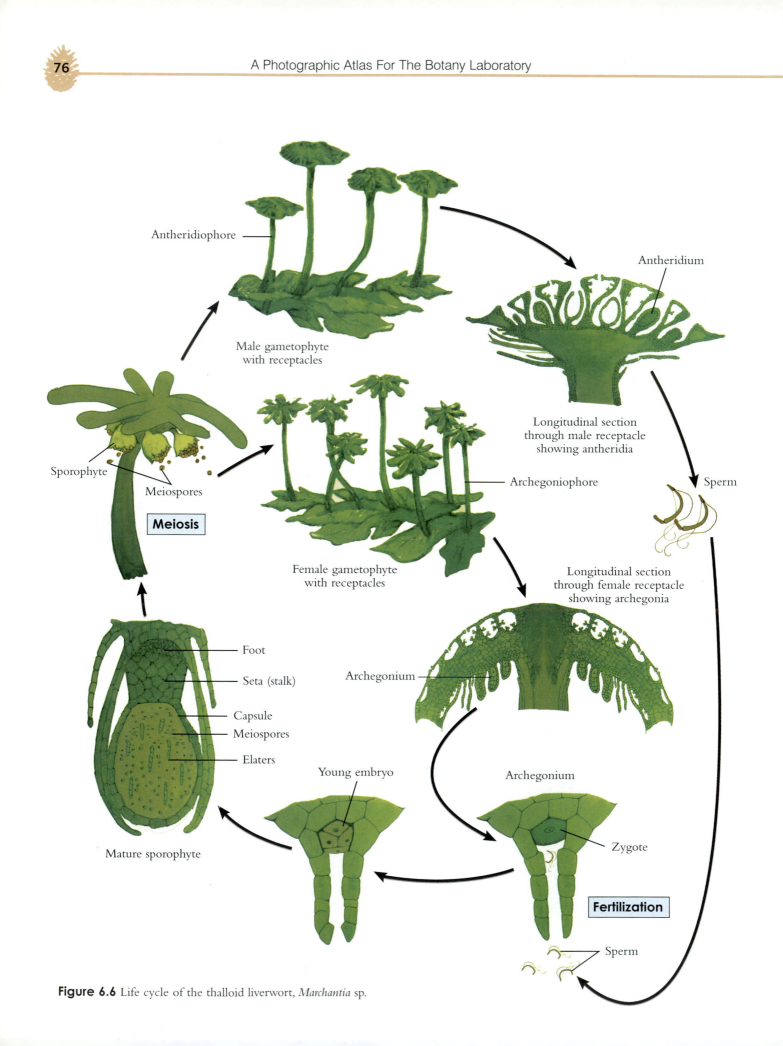

Antheridiophore

Antheridium

Male gametophyte
with receptacles

Longitudinal section
through male receptacle
showing antheridia

Sporophyte

Meiospores

Archegoniophore

Sperm

Meiosis

Female gametophyte
with receptacles

Longitudinal section
through female receptacle
showing archegonia

Foot

Seta (stalk)

Capsule

Meiospores

Elaters

Archegonium

Young embryo

Archegonium

Zygote

Mature sporophyte

Sperm

Fertilization

Sperm

Figure 6.6 Life cycle of the thalloid liverwort, *Marchantia* sp.

Figure 6.7 *Marchantia* sp. with prominent male antheridial receptacles.
1. Antheridial receptacles   3. Gametophyte thallus
2. Antheridiophore

Figure 6.8 Liverwort *Marchantia* sp., showing archegonial receptacles.
1. Archegonial receptacles
2. Archegoniophore

Figure 6.9 *Marchantia* sp. gametophyte plants with prominent gemmae cupules.
1. Gemmae cup with gemmae

Figure 6.10 Transverse section through a gemma cupule of *Marchantia* sp.
1. Gemmae cupule   2. Gemmae

Figure 6.11 Liverwort *Marchantia* sp. showing rhizoids.
1. Rhizoids

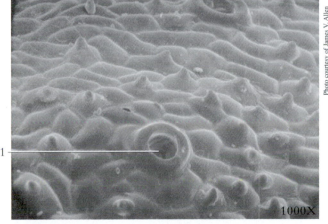

Photo courtesy of James V. Allen

1000X

Figure 6.12 Scanning electron micrograph of the upper surface of the thallus of *Marchantia* sp.
1. Air pore

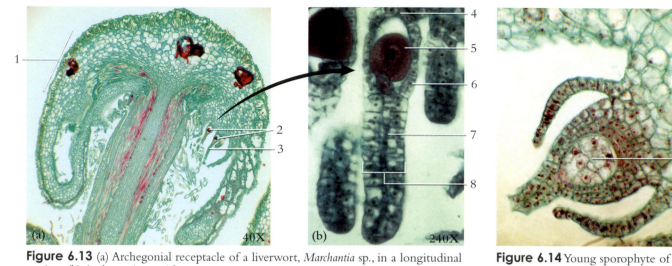

(a) 40X   (b) 240X

Figure 6.13 (a) Archegonial receptacle of a liverwort, *Marchantia* sp., in a longitudinal section. (b) Archegonium with egg.

1. Archegonial receptacle
2. Eggs
3. Neck of archegonium
4. Base of archegonium
5. Egg
6. Venter of archegonium
7. Neck canal
8. Neck of archegonium

Figure 6.14 Young sporophyte of *Marchantia* sp.

1. Young embryo

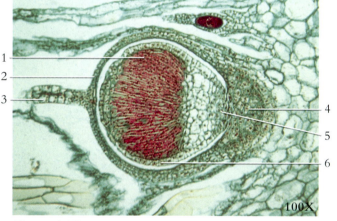

100X

Figure 6.15 Young sporophyte of *Marchantia* sp., in longitudinal section.

1. Sporogenous tissue (*2n*)
2. Enlarged archegonium (calyptra)
3. Neck of archegonium
4. Foot
5. Seta (stalk)
6. Capsule

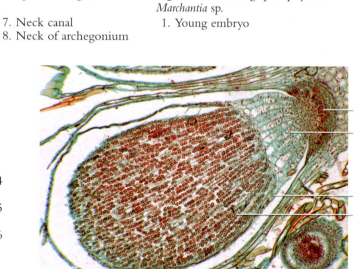

65X

Figure 6.16 Immature and mature sporophytes.

1. Foot
2. Seta (stalk)
3. Sporangium (capsule)
4. Spores (*n*) and elaters (*2n*)

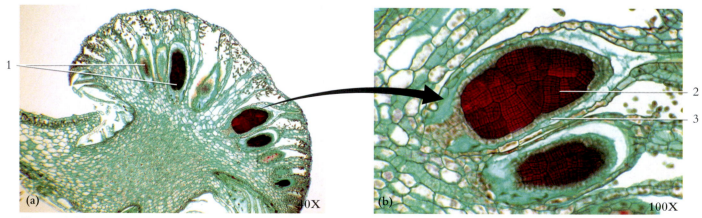

(a) 100X   (b) 100X

Figure 6.17 (a) Male receptacle with antheridia of a liverwort, *Marchantia* sp., in a longitudinal section. (b) Antheridial head showing a developing antheridium.

1. Antheridia
2. Spermatogenous tissue
3. Antheridium

Female receptacle

Sporophyte

Gametophyte

(a)

Sporophyte

Gametophyte

(b)

Figure 6.18 Comparison of the sporophytes and gametophytes of (a) the liverwort, *Marchantia* sp., and (b) the hornwort, *Anthoceros* sp.

Phylum Anthocerophyta - hornworts

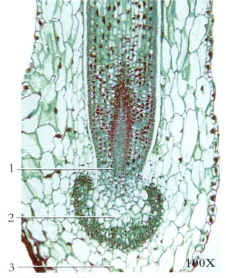

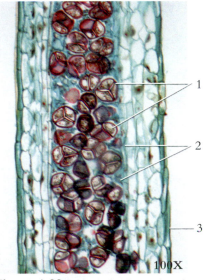

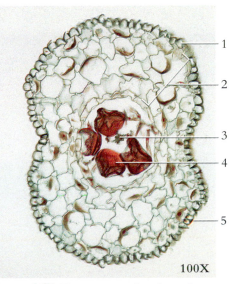

Figure 6.19 Longitudinal section of a portion of the sporophyte of the hornwort, *Anthoceros* sp.

1. Meristematic region of sporophyte
2. Foot
3. Gametophyte

Figure 6.20 Longitudinal section of the sporangium of a sporophyte from the hornwort, *Anthoceros* sp.

1. Spore tetrads
2. Elater-like structures (pseudoelaters)
3. Capsule

Figure 6.21 Transverse section through the capsule of a sporophyte of the hornwort, *Anthoceros* sp.

1. Epidermis
2. Photosynthetic tissue
3. Columella
4. Tetrad of spores
5. Pore (stomate)

Phylum Bryophyta - mosses

Figure 6.22 *Sphagnum* bog in the high Rocky Mountains. This lake has nearly been filled in with dense growths of *Sphagnum* sp.

Figure 6.23 Detail of *Sphagnum* bog showing gametophyte plants.

Figure 6.24 Detail of gametophyte plants of peat moss, *Sphagnum* sp. (scale in mm).

Figure 6.25 Gametophyte plant of peat moss, *Sphagnum* sp., showing attached sporophytes (scale in mm).

1. Sporophyte   3. Gametophyte
2. Pseudopodium

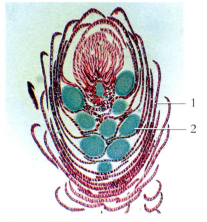

Figure 6.26 Longitudinal section of *Sphagnum* sp. gametophyte showing antheridia.

1. "leaf"   2. Antheridium

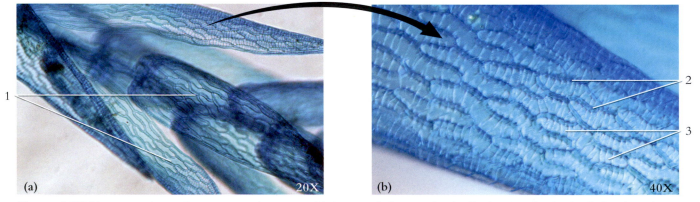

Figure 6.27 (a) Gametophyte of peat moss, *Sphagnum* sp. (b) A magnified view of a "leaf", showing the dead cell chambers that aid in water storage.

1. "Leaves"   2. Photosynthetic cells   3. Dead cells

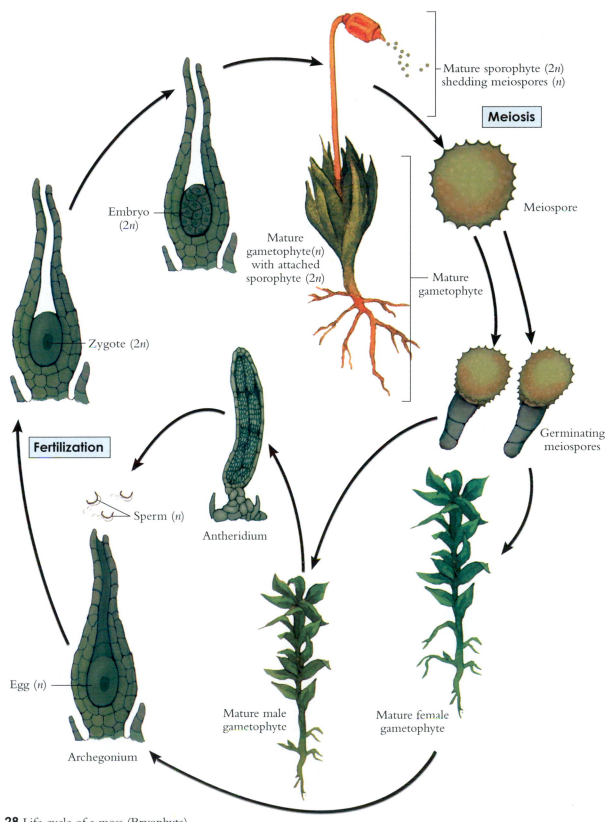

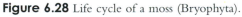

Mature sporophyte (2n)
shedding meiospores (n)

Meiosis

Meiospore

Embryo
(2n)

Mature
gametophyte(n)
with attached
sporophyte (2n)

Mature
gametophyte

Zygote (2n)

Germinating
meiospores

Fertilization

Sperm (n)

Antheridium

Egg (n)

Mature male
gametophyte

Mature female
gametophyte

Archegonium

Figure 6.28 Life cycle of a moss (Bryophyta).

Figure 6.29 Habitat of a moss growing in a wooded environment.
1. Moss
2. Vascular plants

Figure 6.30 Moss–covered sandstone. Under dry conditions, mosses may become dormant and lose their intense green.
1. Stone   2. Moss

Figure 6.31 Four common mosses often used in course work, (a) *Polytrichum* sp., (b) *Mnium* sp., (c) *Hypnum* sp. and (d) *Dicranum* sp.

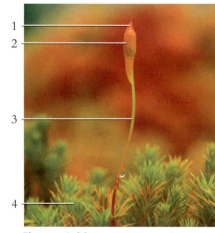

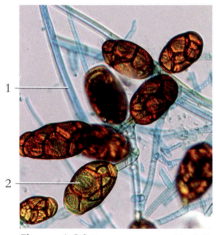

Figure 6.32 Gametophyte plants with sporophyte plant attached.
1. Calyptra
2. Capsule of sporophyte (covered by calyptera)
3. Stalk (seta)
4. Gametophyte

Figure 6.33 Sporophyte plant and capsule.
1. Operculum
2. Capsule of sporophyte (with calyptra absent)
3. Stalk (seta)
4. Gametophyte

Figure 6.34 Protonemata and bulbils of a moss. The bulbils will grow to become a new gametophyte plant.
1. Protonema   2. Bulbil

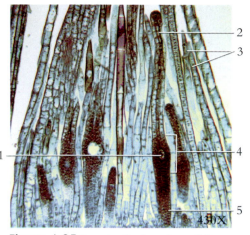

Figure 6.35 Longitudinal section of the archegonial head of the moss *Mnium* sp. The paraphyses are non-reproductive filaments that support the archegonia.

1. Egg   3. Paraphyses   5. Stalk
2. Neck   4. Venter

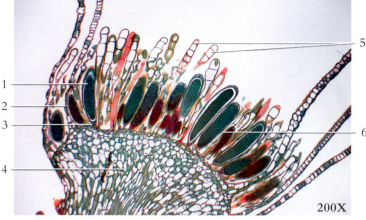

Figure 6.36 Longitudinal section of the antheridial head of the moss *Mnium* sp.

1. Spermatogenous tissue   4. Male gametophyte (*n*)
2. Sterile jacket layer   5. Paraphyses (sterile filaments)
3. Stalk   6. Antheridium (*n*)

Figure 6.37 Close up of *Mnium* sp.

1. Antheridium (*n*)   3. Stalk
2. Spermatogenous tissue   4. Paraphyses

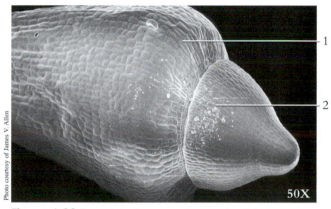

Figure 6.38 Scanning electron micrograph of the sporophyte capsule of the moss *Mnium* sp.

1. Capsule   2. Operculum

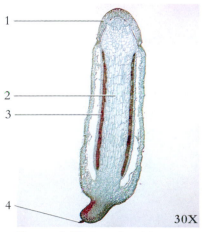

Figure 6.39 Capsule of the moss *Mnium* sp.

1. Operculum   3. Spores
2. Columella   4. Seta

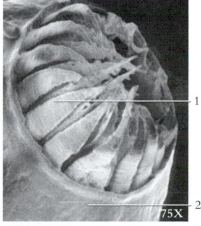

Figure 6.40 Scanning electron micrograph of the peristome of the moss *Mnium* sp. The operculum is absent in the specimen.

1. Peristome   2. Capsule

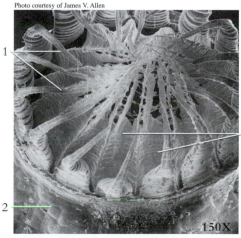

Figure 6.41 Scanning electron micrograph of the peristome of the moss *Mnium* sp.

1. Outer teeth   3. Inner teeth
   of peristome   of peristome
2. Capsule

Chapter 7
Seedless Vascular Plants

Four extant phyla of vascular plants do not produce seeds. These are: Psilotophyta (=Psilophyta)–the whisk ferns; Lycophyta (=Lycopodiophyta)–club mosses and quillworts; Equisetophyta (= Sphenophyta)–horsetails; and Pteridophyta (=Polypodiophyta)–ferns (Table 7.1). Recent evidence from DNA analysis indicates the wisk ferns, horse tails, and ferns are closely related and some botanists place them in a single group, the monilophytes. These plants often inhabit damp, shady environments and are found worldwide. Such plants were particularly abundant during the Carboniferous Period, 290 to 350 million years ago. At that time, many species were large and treelike. Their remains, along with other plants, were compacted to form coal beds. Several thousand species of seedless vascular plants are extant.

Psilopsids are homosporous vascular plants represented by only two living genera, *Psilotum* and *Tmesipteris*, and several species. *Psilotum* is a tropical or subtropical plant. In the United States it occurs in Arizona, Texas, Louisiana, Florida, and Hawaii. It grows readily in greenhouses and may be considered a greenhouse weed. *Tmesipteris* occurs in Australia and islands of the South Pacific, including New Zealand and New Caledonia. Both genera have simple sporophytes, and have no histological distinction between the rhizomes and stems (subterranean and aerial). *Tmesipteris* is an epiphyte on tree ferns and other plants.

Lycophytes are represented by five living genera and about 1,000 species. The most familiar lycophyte is the club moss, *Lycopodium*. The nearly 200 species of this genus are found worldwide in tropical to Arctic regions. Lycophytes are all homosporous. Species of the closely-related *Selaginella* and *Isoetes* are heterosporous.

The 15 living species of horsetails are included in the single genus, *Equisetum*. Three species of *Equisetum* are tropical, and eleven occur in the United States and Canada. Horsetails have jointed stems and scale-like leaves. The stems are further characterized by prominent nodes and elevated siliceous ribs. Horsetails are homosporous, and sporangia develop in a cone at the apex of the stem.

About 12,000 living species of ferns are known. Many have large, "feathery" leaves, called *fronds*. Both roots and fronds grow out of an underground *rhizome*. As fronds develop, they appear to be rolled-up and hence are called *fiddleheads*. Spores produced by sporangia on the underside of the frond of a fern are dispersed by the wind to suitable, moist habitats for germination. Spores germinate, to become a small gametophyte, or *prothallus*. A gametophyte has antheridia that produce sperm and archegonia that produce eggs. Spiral-shaped sperm swim from the antheridia to an archegonium, where fertilization occurs. A zygote forms within the archegonium and becomes an embryo. The young sporophyte soon grows through the gametophyte, takes root, and becomes a mature sporophyte fern plant.

Figure 7.1 Button fern, *Pellaea rotundifolia*.

Table 7.1 The Seedless Vascular Plants

Phyla and Representative Kinds	Characteristics
Phylum Psilotophyta — whisk ferns	Plants small; true roots and leaves absent; rhizome and rhizoids present; homosporous
Phylum Lycophyta (=Lycopodiophyta) — club mosses, quillworts, and spike mosses	Homosporous and heterosporous; many are epiphytes mostly tropical, but some in temperate climates
Phylum Equisetophyta (= Sphenophyta) — horsetails	Epidermis embedded with silica; tips of stems bear cone-like structures containing sporangia; homosporous
Phylum Pteridophyta (=Polypodiophyta) — ferns	Plants often large and conspicuous; leaves complex, rhizome common; mostly homosporous, a few heterosporous

Phylum Psilotophyta - whisk ferns

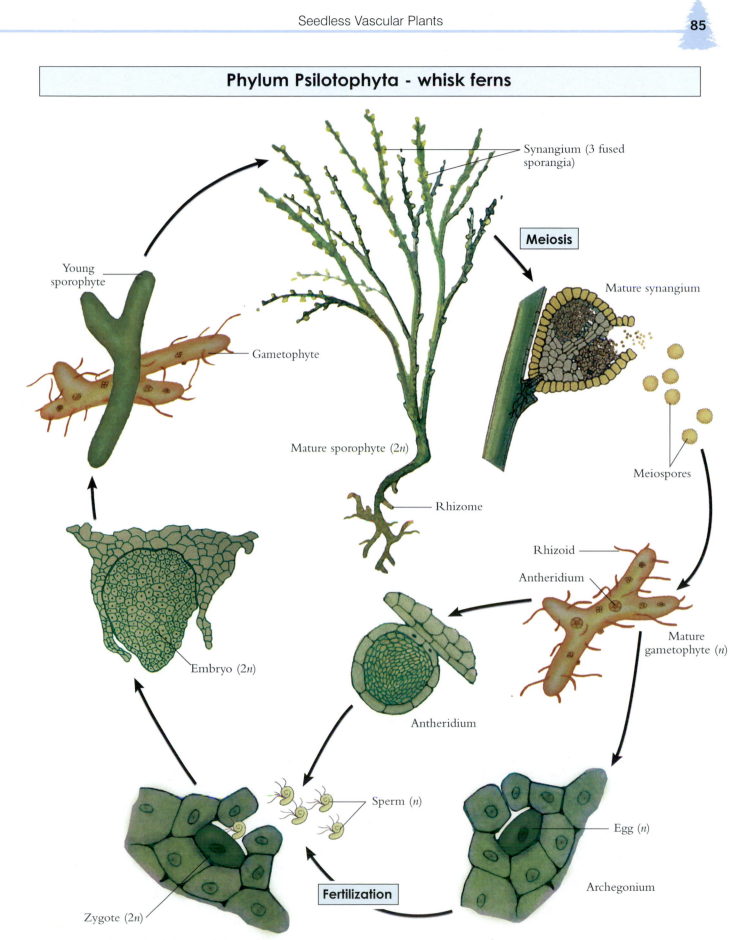

Figure 7.2 Life cycle of the whisk fern, *Psilotum* sp.

Figure 7.3 *Tmesipteris* sp., growing as an epiphyte on a tree fern in Australia.

Figure 7.4 Whisk fern, *Psilotum nudum,* is a simple vascular plant lacking true leaves and roots.

Figure 7.5 Branches (axes) of *Psilotum nudum* (scale in mm)   .
  1. Aerial axis   2. Rhizome

Figure 7.6 Sporophyte of the whisk fern, *Psilotum nudum.* The axes of the sporophyte support sporangia (synangia), which produce spores (scale in mm).
  1. Branch (axis)   2. Sporangia (synangia)

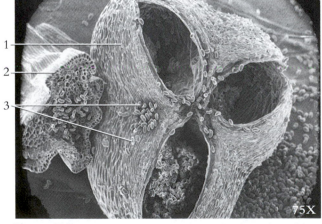

Figure 7.7 Scanning electron micrograph of a ruptured synangium (3 fused sporangia) of *Psilotum* sp., which is spilling spores.
  1. Sporangium (often   2. Axis   3. Spores
     called synangia)

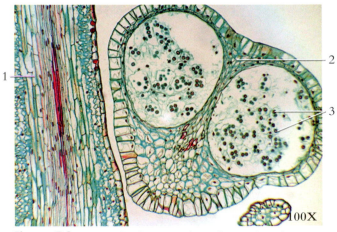

Figure 7.8 Longitudinal section through an axis and sporangium (synangium) of *Psilotum* sp.
  1. Axis   3. Spores
  2. Sporangia (synangium)

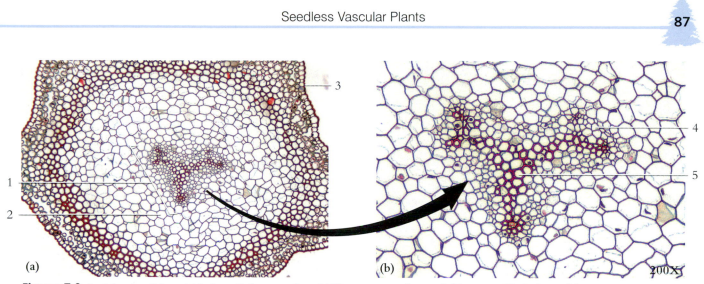

Figure 7.9 Aerial axis of the whisk fern, *Psilotum nudum*. (a) Transverse section and (b) a magnified view of the vascular cylinder (stele).

1. Stele	2. Cortex	3. Epidermis	4. Phloem	5. Xylem

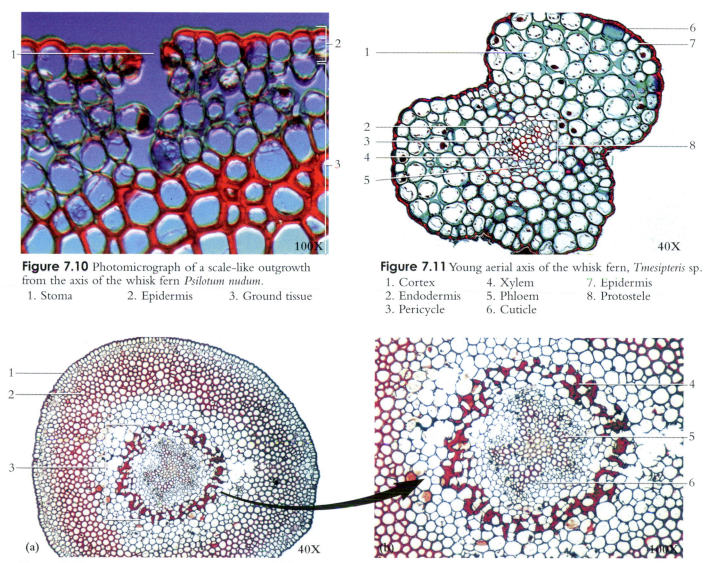

Figure 7.10 Photomicrograph of a scale-like outgrowth from the axis of the whisk fern *Psilotum nudum*.

1. Stoma   2. Epidermis   3. Ground tissue

Figure 7.11 Young aerial axis of the whisk fern, *Tmesipteris* sp.

1. Cortex	4. Xylem	7. Epidermis
2. Endodermis	5. Phloem	8. Protostele
3. Pericycle	6. Cuticle	

Figure 7.12 Older aerial axis of the whisk fern, *Tmesipteris* sp. The genus *Tmesipteris* is restricted to distribution in Australia, New Zealand, New Caledonia, and other South Pacific islands. (a) Axis arising from the aerial axis and (b) a magnified view of the stele.

1. Epidermis	2. Cortex	3. Stele	4. Endodermis	5. Xylem	6. Phloem

Lycophyta (=Lycopodiophyta)- club mosses, quillworts, and spike mosses

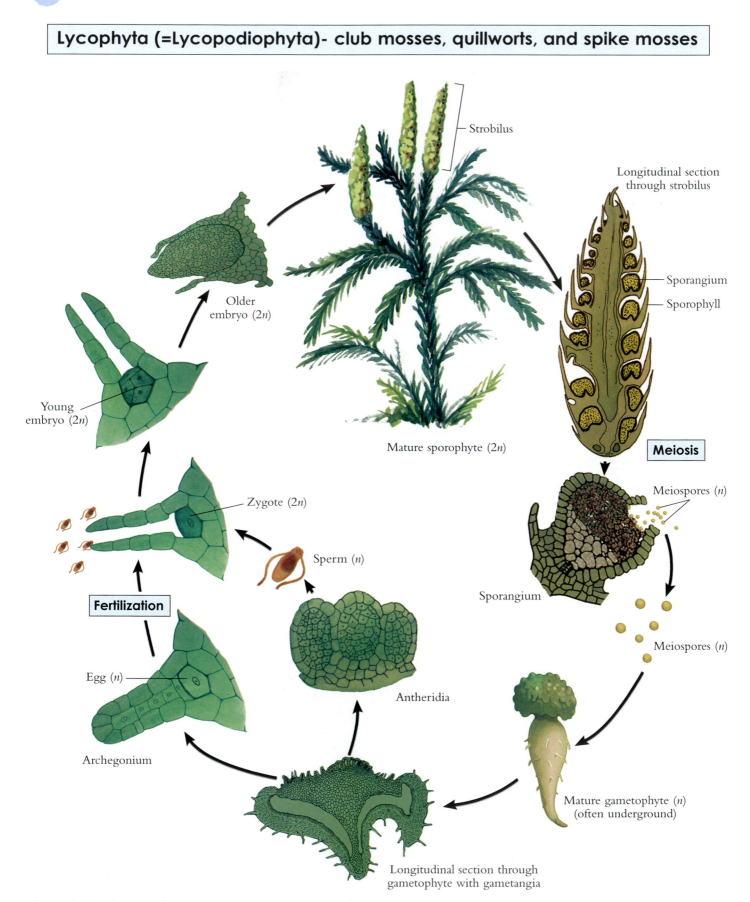

Strobilus

Longitudinal section through strobilus

Older embryo (2n)

Sporangium

Sporophyll

Young embryo (2n)

Mature sporophyte (2n)

Meiosis

Zygote (2n)

Meiospores (n)

Sperm (n)

Sporangium

Fertilization

Meiospores (n)

Egg (n)

Antheridia

Archegonium

Mature gametophyte (n) (often underground)

Longitudinal section through gametophyte with gametangia

Figure 7.13 Life cycle of the homosporous club moss, *Lycopodium* sp.

Figure 7.14 Specimen of a lycopod, *Lycopodium clavatum*, (a) plant and (b) strobilus. *Lycopodium* occurs from the arctic to the tropics (scale in mm).

1. Strobilus   2. Stem

Figure 7.15 Enlargement of a specimen of *Lycopodium* sp., showing branch tip with sporangia on the upper surface of sporophylls (scale in mm).

1. Sporangia
2. Sporophylls (leaves with attached sporangia)

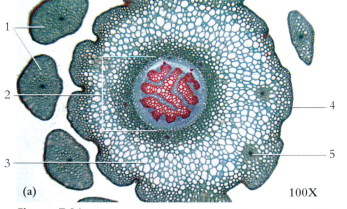

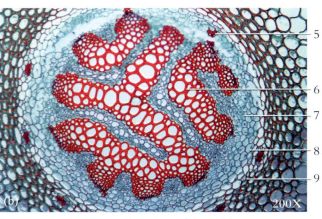

Figure 7.16 (a) Transverse view of an aerial stem of the club moss, *Lycopodium* sp. (b) A magnified view of the stele.

1. Leaves (microphylls)   4. Cortex   5. Leaf trace   7. Phloem   9. Endodermis
2. Stele   3. Epidermis   6. Xylem   8. Pericycle

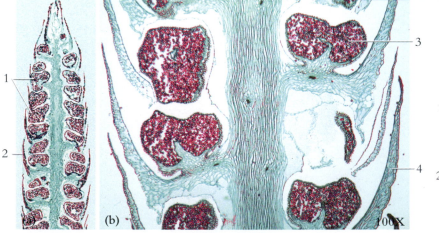

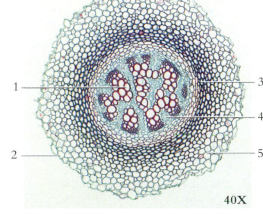

Figure 7.17 (a) Longitudinal section of the strobilus (cone) of the club moss *Lycopodium* sp., and (b) a magnified view of the strobilus showing sporangia.

1. Sporangia   3. Sporangium
2. Sporophyll   4. Sporophyll

Figure 7.18 Transverse section of a rhizome of *Lycopodium* sp. The rhizome of *Lycopodium* is similar to an aerial stem, but it lacks the microphylls.

1. Xylem   3. Endodermis   5. Cortex
2. Epidermis   4. Phloem

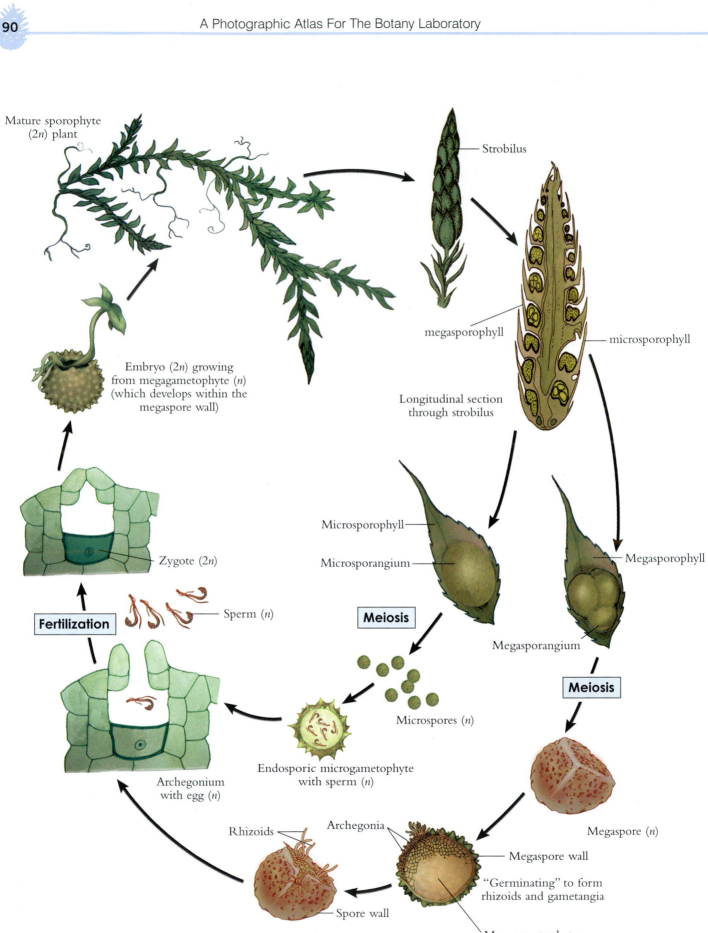

Mature sporophyte (2n) plant

Strobilus

megasporophyll

microsporophyll

Longitudinal section through strobilus

Embryo (2n) growing from megagametophyte (n) (which develops within the megaspore wall)

Microsporophyll

Microsporangium

Megasporophyll

Zygote (2n)

Meiosis

Megasporangium

Meiosis

Fertilization

Sperm (n)

Microspores (n)

Megaspore (n)

Megaspore wall

Endosporic microgametophyte with sperm (n)

Archegonium with egg (n)

Archegonia

"Germinating" to form rhizoids and gametangia

Rhizoids

Spore wall

Megagametophyte

Figure 7.19 Life cycle of *Selaginella* sp. which is heterosporous.

Figure 7.20 Spike moss, *Selaginella kraussiana* (a) growth habit and (b) strobili (cones).
1. Strobili (cones)   2. Sporaphyll with sporangium

Figure 7.21 Spike moss, *Selaginella pulcherrima*.

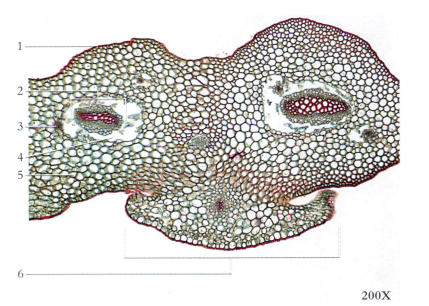

200X

Figure 7.22 Transverse section through stem of *Selaginella* sp. immediately above dichotomous branching.

1. Epidermis
2. Protostele (surrounded by endodermis)
3. Root trace
4. Air cavity
5. Cortex
6. Leaf base

100X

Figure 7.23 Longitudinal section through the strobilus of *Selaginella* sp.

1. Ligule
2. Megasporophyll
3. Megasporangium
4. Megaspore
5. Microsporophyll
6. Microsporangium
7. Microspore
8. Cone axis

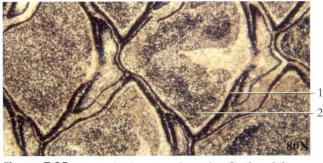

20X

Figure 7.24 Longitudinal view of the surface of the fossil lycophyte *Lepidodendron* sp., a common lycopod from perhaps 300 million years ago.

80X

Figure 7.25 Longitudinal section through a fossil strobilus of the lycophyte *Lepidostrobus* sp., from approximately 300 million years ago.

1. Sporangium   2. Sporogenous tissue

Phylum Sphenophyta (=Equisetophyta) - horsetail

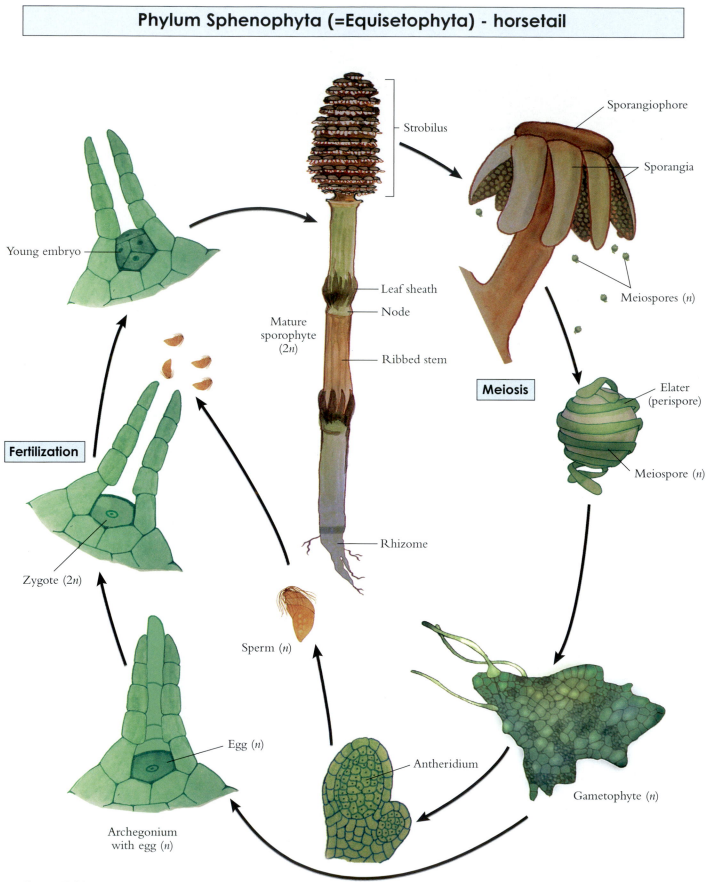

Figure 7.26 Life cycle of the horsetail, *Equisetum* sp.

Figure 7.27 *Equisetum telmateia* showing lateral branching.

Figure 7.28 Close-up of *Equisetum telmateia* showing lateral branches growing through leaf sheath.

Figure 7.29 Stems of *Equisetum* sp. without lateral branching and showing a prominent leaf sheath at the node.
1. Stem   2. Leaf sheath

Figure 7.30 Horsetail, *Equisetum* sp. Numerous species of Equisetophyta were abundant throughout tropical regions during the Paleozoic Era, some 300 million years ago. Currently, Equisetophyta is represented by this single genus. The meadow horsetail, *Equisetum* sp., showing (a) an immature strobilus, (b) mature strobilus, shedding spores, (c) an open strobilus, and (d) a sporangiophore with its spores released.
1. Sporangiophores
2. Separated sporangiophores revealing sporangia
3. Sporangiophores after spores are shed
4. Open sporangia with spores shed

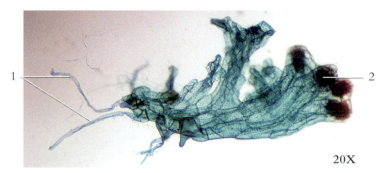

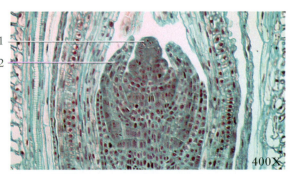

Figure 7.31 *Equisetum* sp., young gametophyte.
1. Rhizoids   2. Antheridium

Figure 7.32 Longitudinal section of *Equisetum* sp. shoot apex.
1. Apical cell   2. Leaf primordium

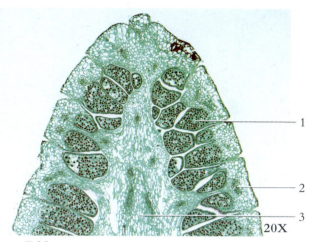

Figure 7.33 Longitudinal section of *Equisetum* sp. strobilus.
1. Sporangium   2. Sporangiophore   3. Strobilus axis

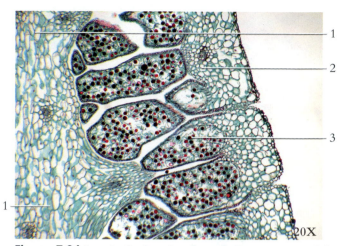

Figure 7.34 Longitudinal section through *Equisetum* sp. strobilus.
1. Axis of the strobilus   2. Sporangiophore   3. Sporangium

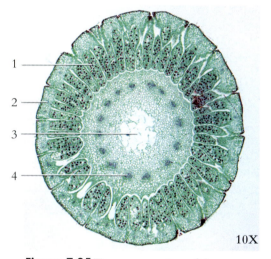

Figure 7.35 Transverse section of the strobilus of *Equisetum* sp.
1. Sporangium   3. Strobilus axis
2. Sporangiophore   4. Vascular bundle

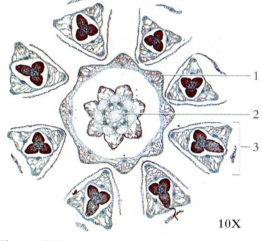

Figure 7.36 Transverse section of the stem of *Equisetum* sp. just above a node.
1. Leaf sheath   2. Main stem   3. Branch

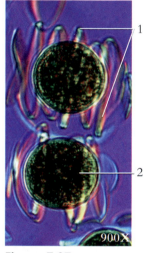

Figure 7.37 Meiospores of *Equisetum* sp.
1. Perispore (elater)
2. Meiospore

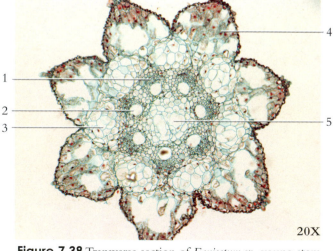

Figure 7.38 Transverse section of *Equisetum* sp. young stem.
1. Vascular tissue   3. Future air canal   5. Pith
2. Air canal   4. Cortex

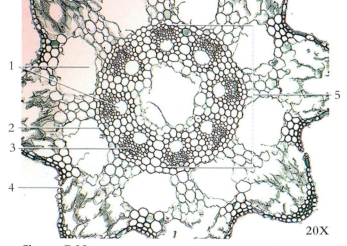

Figure 7.39 Transverse section of *Equisetum* sp. older stem.
1. Air canals   3. Vascular tissue   5. Eustele
2. Endodermis   4. Stomate

Phylum Pteridophyta (=Polypodiophyta) - ferns

Figure 7.40 Life cycle of a fern.

Figure 7.41 Water fern, *Azolla* sp., is a floating fresh-water plant found throughout Europe and the United States. *Azolla* may become bright orange or red during fall.

(a)   (b)

Figure 7.42 View of a new (a) compound and (b) simple fern leaf showing circinate vernation forming a fiddlehead.

Figure 7.43 Fronds of the staghorn fern, *Platycerium alcicorne*.

Figure 7.44 Pinnate leaf showing dichotomous venation in the leaflets of a fern.

  1. Leaf   2. Pinnae   3. Venation

Figure 7.45 Leaf of the fern *Phanerophlebia* sp., or holly fern.

Figure 7.46 Leaf of the fern *Phanerophlebia* sp., showing sori (groups of sporangia).
  1. Pinna
  2. Sori

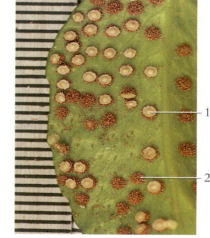

Figure 7.47 Close-up of the fern leaf of *Phanerophlebia* sp. (scale in mm).
  1. Sorus with indusium
  2. Sorus with indusium shed

Figure 7.48 Leaf of the fern *Polypodium virginianum*.

Figure 7.49 Leaf of the fern *Polypodium virginianum*, showing sori (groups of sporangia).
1. Pinna   2. Sori

Figure 7.50 Close-up of the fern pinna of *Polypodium virginianum* (scale in mm).
1. Sorus

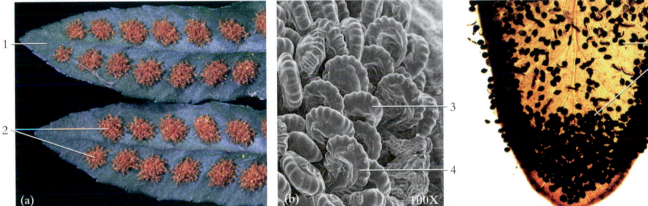

Figure 7.51 Fern *Polypodium* sp. (a) Sori on the undersurface of the pinnae, and (b) a scanning electron micrograph of a sorus.
1. Pinna   2. Sori   3. Annulus   4. Sporangium

Figure 7.52 Magnified view of the fern pinna of *Pteridium* sp. showing numerous scattered sporangia.
1. Sporangia

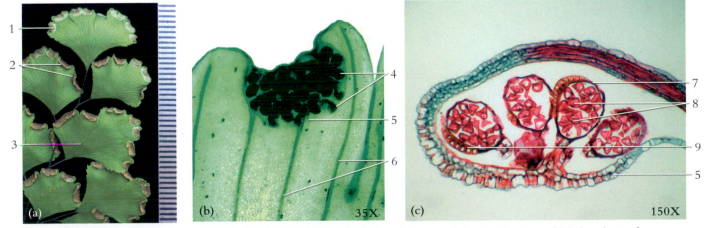

Figure 7.53 Maidenhair fern *Adiantum* sp. (a) Pinnae and sori. (b) Magnified view of the tip of a pinna folded under to form a false indusium that encloses the sorus. (c) Sorus with sporangia containing spores (scale in mm).
1. False indusium   3. Pinna   5. False indusium enfolding a sorus   7. Sporangium   9. Annulus
2. Sori   4. Sporangia with spores   6. Vascular tissue (veins) of the pinna   8. Spores

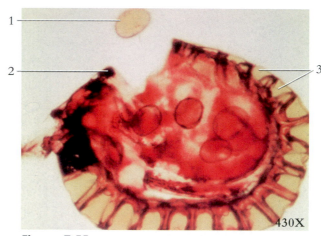

Figure 7.54 Sorus of a fern, *Cyrtomium falcatum*.
1. Sporangium (*2n*)   3. Indusium (*2n*)   5. Spores (*n*)
2. Annulus   4. Pinna tissue

Figure 7.55 Sporangium of the fern *Cyrtomium* sp., discharging spores.
1. Spore (*n*)   3. Annulus
2. Lip cell

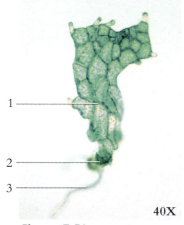

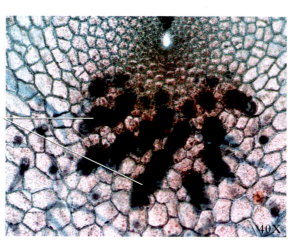

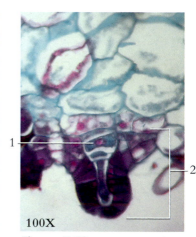

Figure 7.56 Young fern gametophyte.
1. Gametophyte   3. Rhizoid
2. Spore cell wall

Figure 7.57 Fern gametophyte with archegonia.
1. Archegonia

Figure 7.58 Fern gametophyte showing archegonium.
1. Egg
2. Archegonium

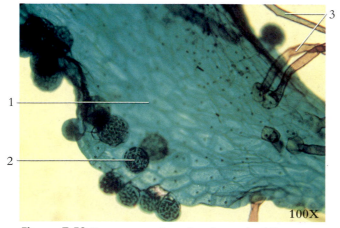

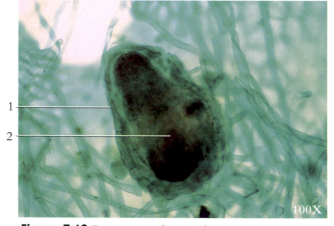

Figure 7.59 Fern gametophyte showing antheridia.
1. Gametophyte (prothallus)   3. Rhizoids
2. Antheridium with sperm

Figure 7.60 Fern gametophyte with a young sporophyte attached.
1. Expanded archegonium   2. Young sporophyte

Figure 7.61 Transverse section through the stem of a fern, *Dicksonia* sp. showing a siphonostele.

1. Phloem   3. Sclerified pith   5. Pith
2. Xylem   4. Cortex

Figure 7.62 Tree fern, *Cyathea medullaris*, a native to the North Island of New Zealand.

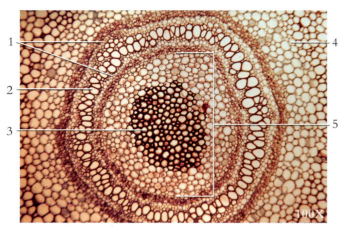

Matonidium brownii

Astralopteris sp.

Osmundacaulis sp.

Figure 7.63 Fossil impressions of the leaves (a) and (b), and transverse section of vascular tissue (c) of ferns. These fossils are approximately 65 to 70 million years old

Figure 7.64 Heterosporous water fern, *Salvinia* sp. (scale in mm.).

Figure 7.65 Heterosporous water fern, *Salvinia molesta*, a highly invasive plant clogging waterways in warm climates.

Figure 7.67 Water fern, *Marsilea* sp., growing in a pond.

Figure 7.68 Water fern, *Marsilea* sp., showing detail of pinnae.

Figure 7.69 Water fern, *Marsilea* sp., fiddlehead.

Figure 7.70 Water fern, *Marsilea* sp., sporocarp.

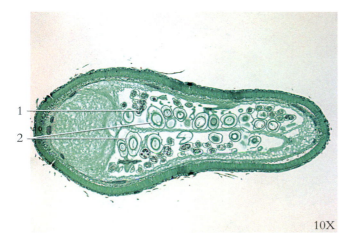

10X

Figure 7.71 Transverse section of a sporocarp of the water fern, *Marsilea* sp., which is one of the two living orders of heterosporous ferns.
  1. Microsporangium with microspores
  2. Megasporangia with megaspores

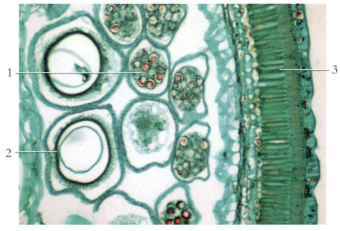

Figure 7.72 Magnified view transverse section of a sporocarp of *Marsilea* sp.
  1. Microsporangium with microspores
  2. Megasporangium showing 1 megaspore
  3. Sporocarp wall

Chapter 8
Gymnosperms: Exposed Seed Plants

Gymnosperms (exposed seed plants) include plants in four phyla: Cycadophyta (cycads); Ginkgophyta (ginkgos); Coniferophyta (=Pinophyta) (conifers); and Gnetophyta (gnetophytes) (Table 8.1). Gymnosperms arose in the late Devonian period, about 375 to 400 million years ago. They dominated land floras throughout most of the Mesozoic Era until the late Cretaceous Period, some 70 million years ago. Currently, about 65 genera and 720 extant species are known.

Reproduction in seed plants is well adapted to a land existence. Seeds develop from protective structures known as *ovules*, which mature to produce protective and nutritive layers around the embryo. Gymnosperms produce seeds in protective *cones*. In the life cycle of a gymnosperm, such as a pine, the mature *sporophyte* (tree) has cones that produce megaspores that develop into female gametophyte generations, and cones that produce microspores that develop into male gametophyte generations—mature pollen grains. Following fertilization, immature sporophyte generations are present in seeds located in the female cones. The cone opens at maturity or following an environmental signal and the seeds disperse and germinate if conditions are favorable. Reproduction in flowering plants (chapter 9) is similar to gymnosperms except that angiosperm pollen and ovules are produced in flowers, rather than in cones and seeds develop within protective fruits, and the life cycle is faster.

Cycads are gymnosperms superficially resembling palm trees. They are represented by 10 extant genera and about 100 species. They are all tropical or subtropical in distribution. Ovules and seeds of cycads are exposed, and sperm are flagellated and motile. Most cycads have short stems, with large pinnate leaves attached at the crown. The leaves are thick with sunken stomata. Most cycads are less than 3 meters tall, although, one species reaches 20 meters in height.

A single living species, *Ginkgo biloba*, comprises the phylum Ginkgophyta. *Ginkgo biloba* is native to China, but has been cultivated extensively throughout the world in temperate climates. Ginkgophyta is an ancient phylum in existence since late Carboniferous times, some 290 million years ago. The *Ginkgo*, or maidenhair tree, grows to 35 meters tall. The leaves of ginkgos are fan-shaped, with long petioles. Because it is slow-growing and because the wood is brittle, *Ginkgo* is not a good source of lumber.

Conifers are mostly evergreen woody shrubs and trees. About 50 genera, with perhaps 550 living species comprise Pinophyta (=Coniferophyta). Conifers lack flowers and their seeds are exposed on the surface of cone scales. Conifers often have needle-like leaves and stomata are sunken. Conifers supply much of the lumber for building and wood for the manufacture of paper, turpentine, and many other products. Conifers are also a major source of wood for fuel.

Three genera of gnetophytes exist—*Gnetum, Ephedra,* and *Welwitschia*—with about 70 living species. An interesting feature of gnetophytes is the presence of vessel elements in their xylem tissue. Most flowering plants contain vessels elements as well, providing evidence to some botanists Gnetophyta might have been ancestral to angiosperms. Each of the three genera of gnetophytes is very different. *Gnetum* is a vine or small tree resembling a flowering plant in physical appearance. It occurs in tropical rain forests. *Ephedra* is a highly branched shrub occurring in dry or arid habitats including the southwest deserts of the United States. *Welwitschia* does not resemble any other plant on earth. Found in Africa, it has a short, thick, disk-shaped stem with two long leaves attached at its edge.

Table 8.1 Gymnosperms within the Kingdom Plantae

Phyla and Representative Kinds	Characteristics
Phylum Cycadophyta— cycads	Gymnosperms with pollen and seed cones borne on different plants; motile sperm; plants mostly with single short, stout stems or small trees; leaves large, palm-like
Phylum Ginkgophyta— *Ginkgo*	Gymnosperm with deciduous, fan-shaped leaves; motile sperm; large tree
Phylum Pinophyta (= Coniferophyta)— conifers	Woody gymnosperms, producing seeds in cones; motile sperm absent; most with needle-like leaves; stoma sunken; often large trees
Phylum Gnetophyta— gnetophytes	Gymnosperms that contain vessel elements; motile sperm absent; mostly shrub-like

Phylum Cycadophyta - cycads

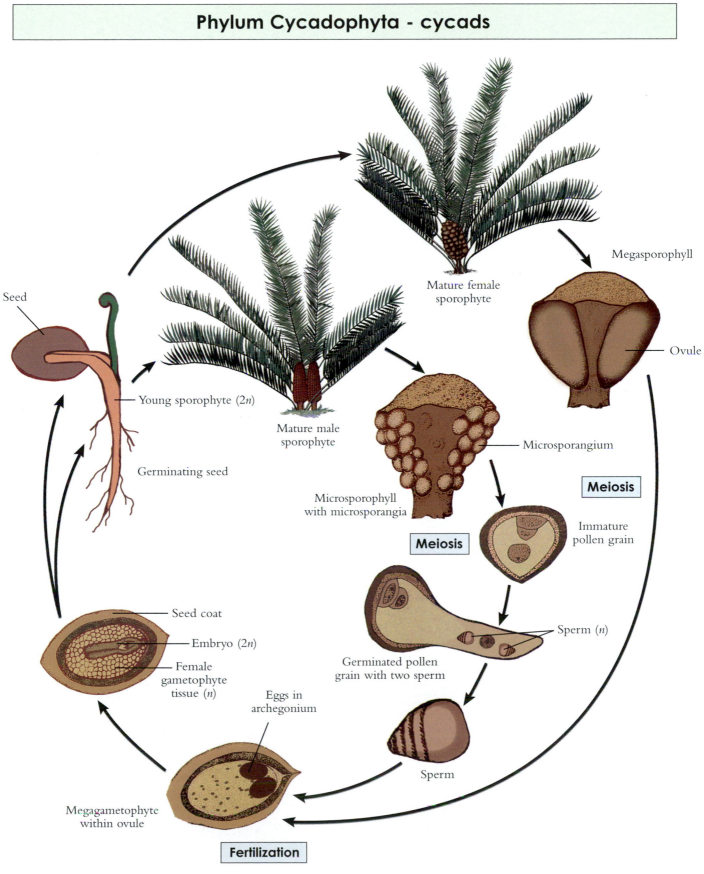

Figure 8.1 Life cycle of a cycad.

Figure 8.2 *Cycas revoluta*. Cycads were abundant during the Mesozoic Era. Currently, there are 10 living genera, with about 100 species, that are mainly found in tropical and subtropical areas. The trunk of many cycads is densely covered with petioles of shed leaves.

Figure 8.3 *Cycas revoluta* showing a female cone.
1. Cone

Figure 8.4 *Cycas revoluta* showing a close up view of a female cone with developing seeds.
  1. Seeds   2. Megasporophyll

Figure 8.5 *Cycas revoluta* showing a close up view of a female cone during seed dispersal.
  1. Seeds

Figure 8.6 Male cone of *Cycas revoluta*.
1. Cone

Figure 8.7 Male cone of *Cycas revoluta* after release of pollen.

Figure 8.8 Young plant of the cycad *Zamia pumila*. Found in Florida, this cycad is the only species native to the United States. The rootstocks and stems of this plant were an important source of food for some Native Americans.

Figure 8.9 Microsporangiate cones of the cycad *Zamia* sp.

Figure 8.10 *Encephalartos villosus*, a non-threatened species of cycad native to southeastern Africa.

Figure 8.11 Maturing female cone of *Encephalartos villosus*.

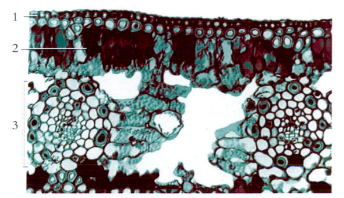

Figure 8.12 Transverse section of the leaf of the cycad *Zamia* sp.
1. Upper epidermis
2. Palisade mesophyll
3. Vascular bundle (vein)

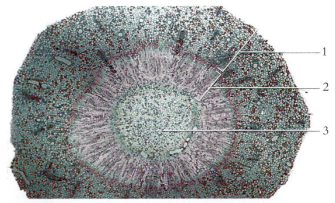

Figure 8.13 Transverse section of the stem of the cycad *Zamia* sp.
1. Cortex   3. Pith
2. Vascular tissue

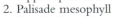

Figure 8.14 (a) Microsporangiate cone of the cycad, *Zamia* sp. The cone on the right (b) is longitudinally sectioned.
  1. Microsporangia   2. Microsporophyll

Figure 8.15 Microsporangiate cone of a cycad showing microsporangia on microsporophylls.
  1. Microsporangium   2. Microsporophyll

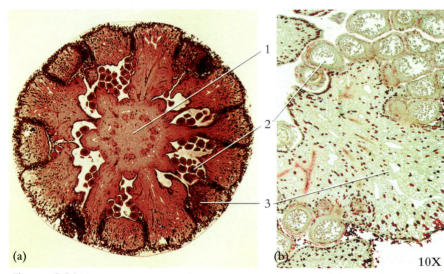

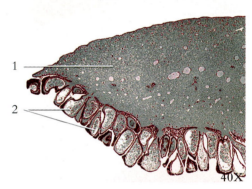

Figure 8.17 Longitudinal section of a microsporophyll of the cycad *Cycas* sp. Note the microsporangia develop on the undersurface of the microsporophyll.
  1. Microsporophyll
  2. Microsporangia

Figure 8.16 Transverse sections of a microsporangiate cone of the cycad *Zamia* sp. (a) A low magnification, and (b) a magnified view.
  1. Cone axis   2. Microsporangia   3. Microsporophyll

Figure 8.18 Megasporangiate cone of *Cycas revoluta* showing ovules on leaf-like megasporophylls near the time of pollination.

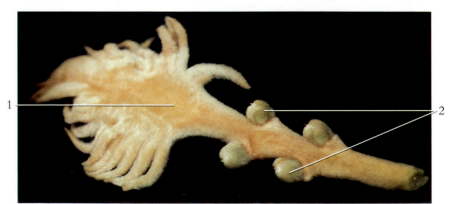

Figure 8.19 Megasporophyll and ovules of *Cycas revoluta*.
  1. Megasporophyll   2. Ovules

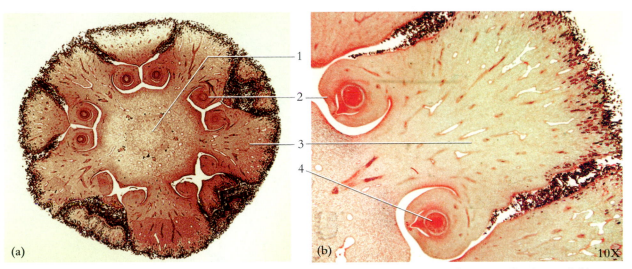

(a) (b) 10X

Figure 8.20 Transverse sections of a megasporangiate cone of the cycad *Zamia* sp. (a) A low magnification, and (b) a magnified view.

1. Cone axis   2. Ovule   3. Megasporophyll   4. Megasporocyte

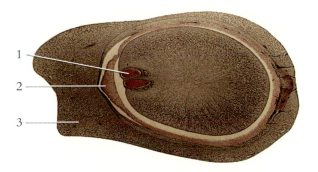

Figure 8.21 Ovule of the cycad *Zamia* sp. The ovule has two archegonia and is ready to be fertilized.

1. Archegonium
2. Megasporangium (nucellus)
3. Integument (will become seed coat)

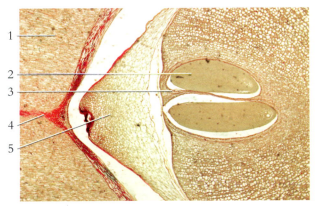

Figure 8.22 Magnified view of the ovule of the cycad *Zamia* sp. showing eggs in archegonia.

1. Integument   4. Micropylearea
2. Egg   5. Megasporangium
3. Archegonium

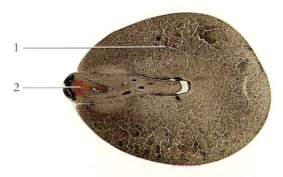

Figure 8.23 Ovule of the cycad *Zamia* sp. The ovule has been fertilized and contains an embryo. The seed coat has been removed from this specimen.

1. Female gametophyte   2. Embryo

Figure 8.24 Magnified view of the ovule of the cycad *Zamia* sp. showing the embryo.

1. Leaf primordium   4. Cotyledon
2. Root apex   5. Female gametophyte
3. Shoot apex

Figure 8.25 *Ginkgo biloba*, or maidenhair tree. Consisting of a central trunk with lateral branches, a mature *Ginkgo* grows to 100 meters tall. Native to China, *Ginkgo biloba* has been introduced into countries with temperate climates throughout the world as an interesting and hardy ornamental tree.

Figure 8.27 Fossil *Ginkgo biloba* leaf impression from Paleocene sediment. This specimen was found in Morton County, North Dakota.

Figure 8.29 Branch of a *Ginkgo biloba* tree supporting a mature seed.

  1. Long shoot   3. Mature seed
  2. Short shoot (spur)

Figure 8.26 Leaf from the *Ginkgo biloba* tree. The fan-shaped leaf is characteristic of this species.

Figure 8.28 As the sole member of the phylum Ginkgophyta, *Ginkgo biloba* is able to withstand air pollution. Ginkgos are often used as ornamental trees within city parks. *Ginkgo biloba* may have the longest genetic lineage among seed plants.

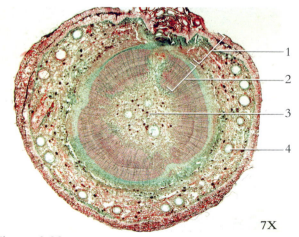

7X

Figure 8.30 Transverse section of a short branch from *Ginkgo biloba*.

  1. Cortex   3. Pith
  2. Vascular tissue   4. Mucilage duct

Figure 8.31 Leaves and immature ovules on a short shoot of the ginkgo tree, *Ginkgo biloba*.

1. Leaf
2. Immature ovules
3. Short shoot
4. Long shoot

Figure 8.32 Pollen strobili of the ginkgo tree, *Ginkgo biloba*.

1. Leaf
2. Long shoot
3. Pollen strobilus

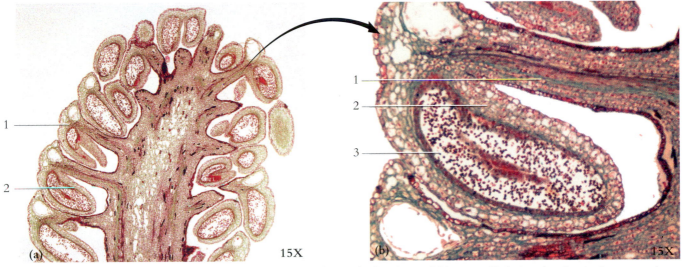

(a)   15X
(b)   15X

Figure 8.33 Microsporangiate strobilus of *Ginkgo biloba*. (a) A longitudinal section and (b) a magnified view showing a microsporangium.

1. Sporophyll
2. Microsporangium
3. Pollen

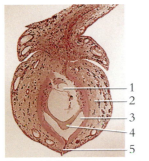

Figure 8.34 Longitudinal section of an ovule of *Ginkgo biloba* prior to fertilization.

1. Megagametophyte
2. Integument
3. Pollen chamber
4. Nucellus
5. Micropyle

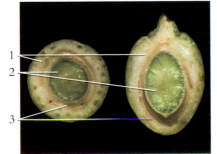

Figure 8.35 Transverse and longitudinal sections through a living immature seed of *Ginkgo biloba* showing the green megagametophyte.

1. Fleshy layer of integument
2. Megagametophyte
3. Stoney layer of integument

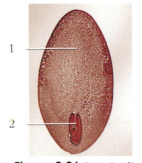

Figure 8.36 Longitudinal section of a seed of *Ginkgo, biloba* with the seed coat removed.

1. Megagametophyte
2. Developing embryo

Figure 8.37 Magnified view of the ovule of *Ginkgo biloba* showing the embryo.

1. Leaf primordium
2. Shoot apex
3. Root apex
4. Megagametophyte

Pinophyta - conifers

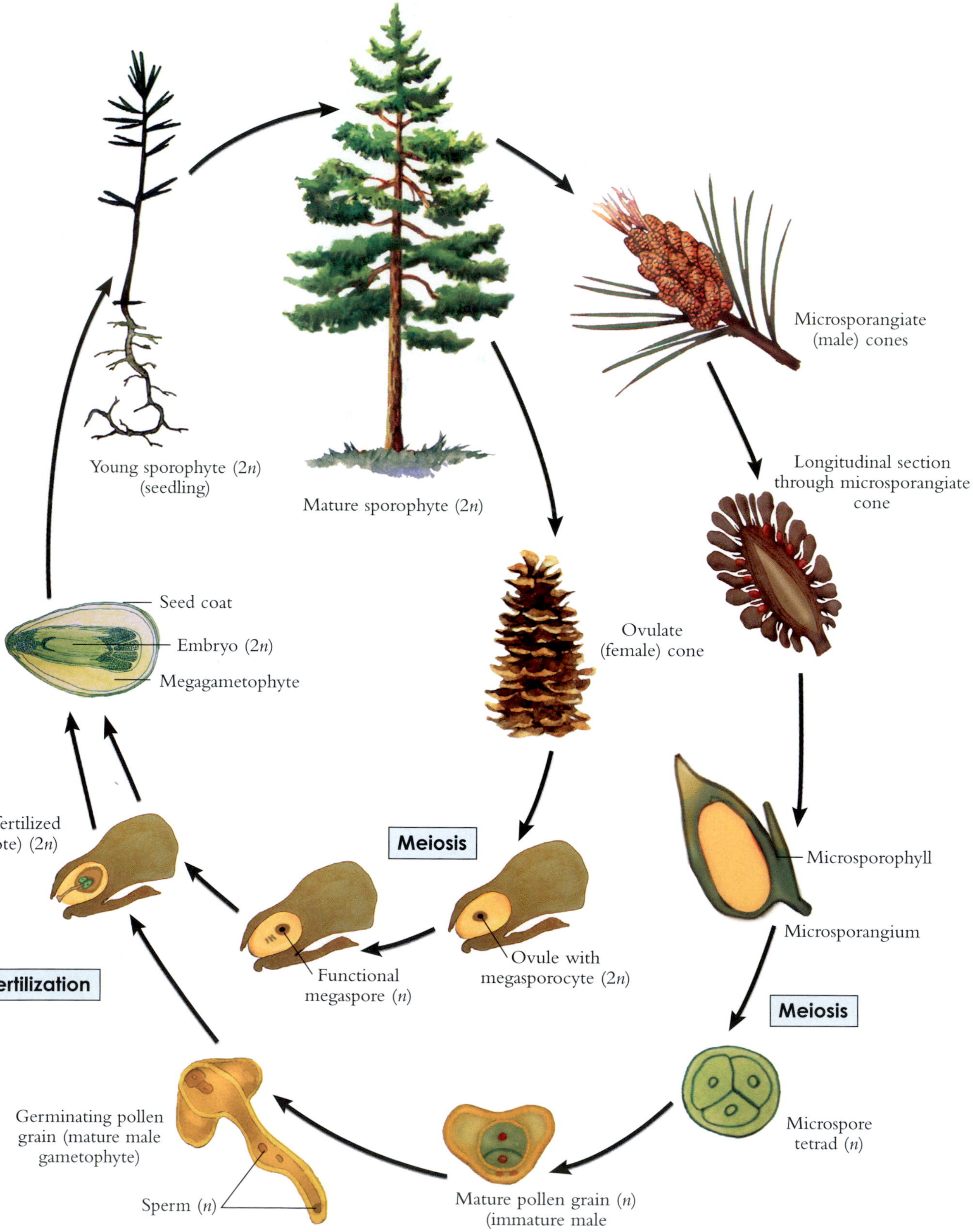

Microsporangiate
(male) cones

Young sporophyte (2n)
(seedling)

Mature sporophyte (2n)

Longitudinal section
through microsporangiate
cone

Seed coat

Embryo (2n)

Megagametophyte

Ovulate
(female) cone

Microsporophyll

Microsporangium

Egg fertilized
(zygote) (2n)

Meiosis

Meiosis

Fertilization

Functional
megaspore (n)

Ovule with
megasporocyte (2n)

Microspore
tetrad (n)

Germinating pollen
grain (mature male
gametophyte)

Sperm (n)

Mature pollen grain (n)
(immature male
gametophyte)

Figure 8.38 Life cycle of the pine, *Pinus* sp.

Courtesy of Champion Paper Company, Inc.

Figure 8.39 Diagram of the tissues in the stem (trunk) of a conifer. The periderm and dead secondary phloem (outer bark) protects the tree against water lost and the infestation of insects and fungi. The cells of the phloem (inner bark) compress and become nonfunctional after a relatively short period. The vascular cambium annually produces new phloem and xylem and accounts for the growth rings in the wood. The secondary xylem is a water transporting layer of the stem and provides structural support to the tree.

1. Outer bark
2. Phloem
3. Vascular cambium
4. Secondary xylem

Courtesy of Champion Paper Company, Inc.

Figure 8.40 Stem (trunk) of a pine tree that was harvested in the year 2000 when the tree was 62 years old. The growth rings of a tree indicate environmental conditions that occurred during the tree's life.

1. **1939**—A pine seedling.
2. **1944**—Healthy, undisturbed growth indicated by broad and evenly spaced rings.
3. **1949**—Growth disparity probably due to the falling of a dead tree onto the young healthy six-year old tree. The wider "reaction rings" on the lower side help support the tree.
4. **1959**—The tree is growing straight again, but the narrow rings indicate competition for sunlight and moisture from neighboring trees.
5. **1962**—The surrounding trees are harvested, thus permitting rapid growth once again.
6. **1965**—A burn scar from a fire that quickly scorched the forest.
7. **1977**—Narrow growth rings resulting from a prolonged drought.
8. **1992**— Narrowgrowth rings, resulting from a sawfly insect infestation, whose larvae eat the needles and buds of many kinds of conifers.

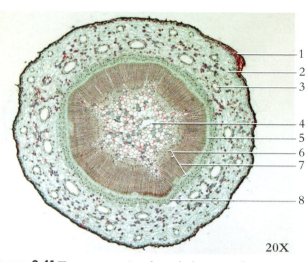

Figure 8.41 Transverse section through the stem of a young conifer showing the arrangement of the tissue layers.

1. Epidermis
2. Cortex
3. Resin duct
4. Pith
5. Cambium
6. Primary xylem
7. Spring wood of secondary xylem
8. Primary phloem

20X

Figure 8.42 Transverse section through the stem of *Pinus* sp., showing secondary stem growth.

1. Bark (cortex, and periderm)
2. Secondary phloem
3. Vascular cambium
4. Pith
5. Secondary xylem
6. Resin duct
7. Epidermis

5X

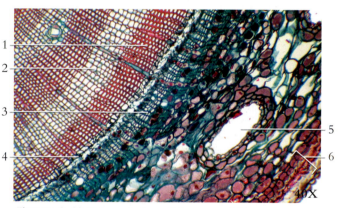

Figure 8.43 Enlarged view of the stem of *Pinus* sp. showing tissues following secondary growth.

1. Late secondary xylem (wood)
2. Early secondary xylem (wood)
3. Secondary phloem
4. Vascular cambium
5. Resin duct
6. Periderm

40X

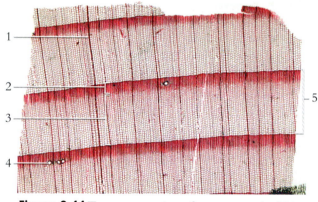

Figure 8.44 Transverse section of mature wood of *Pinus* sp. showing three growth rings.

1. Ray
2. Late secondary xylem (wood)
3. Early secondary xylem (wood)
4. Resin duct
5. One ring

Figure 8.45 Radial longitudinal section through the phloem of *Pinus* sp.

1. Sieve areas on a sieve cell
2. Storage parenchyma
3. Sieve cell

100X

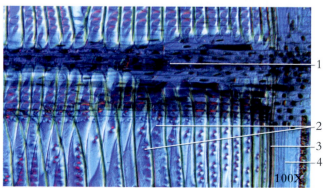

Figure 8.46 Radial longitudinal section through a stem of *Pinus* sp., cut through the xylem tissue.

1. Ray parenchyma
2. Tracheids
3. Vascular cambium
4. Sieve cells

100X

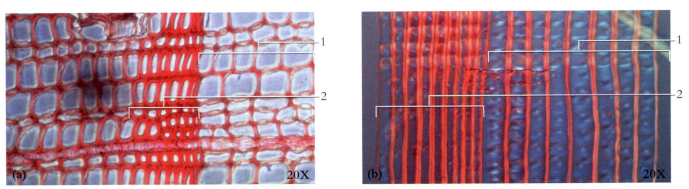

Figure 8.47 Growth rings in *Pinus* sp. (a) Transverse section through a stem; and (b) radial longitudinal section through a stem.

1. Early wood   2. Late wood

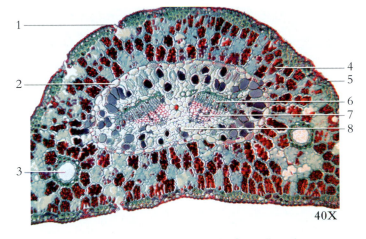

Figure 8.48 Transverse section of a leaf (needle) of *Pinus* sp.

1. Stoma   5. Epidermis
2. Endodermis   6. Phloem
3. Resin duct   7. Xylem
4. Photosynthetic mesophyll   8. Transfusion tissue

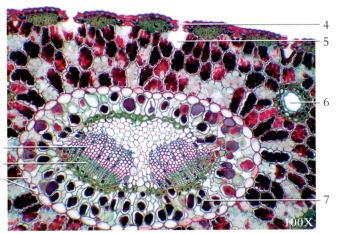

Figure 8.49 Transverse section through the leaf (needle) of *Pinus* sp.

1. Xylem   5. Sub-stomatal chamber
2. Phloem   6. Resin duct
3. Endodermis   7. Transfusion tissue (surrounding
4. Sunken stoma   vascular tissue)

Figure 8.50 Transverse section of a leaf (needle) of *Pinus* sp.

1. Mesophyll cell   4. Epidermis
2. Epithelium   5. Hypodermis
3. Resin duct   6. Tannins in vacuole

Figure 8.51 Transverse section through the leaf (needle) of *Pinus* sp.

1. Endodermis   5. Sunken stoma
2. Casparian strip   6. Hypodermis
3. Mesophyll cell   7. Epidermis
4. Sub-stomatal chamber

Picea sp.

Abies sp.

Pinus sp.

Thuja sp.

Taxodium sp.

Taxus sp.

Figure 8.52 Megasporangiate cones from various species of conifers.

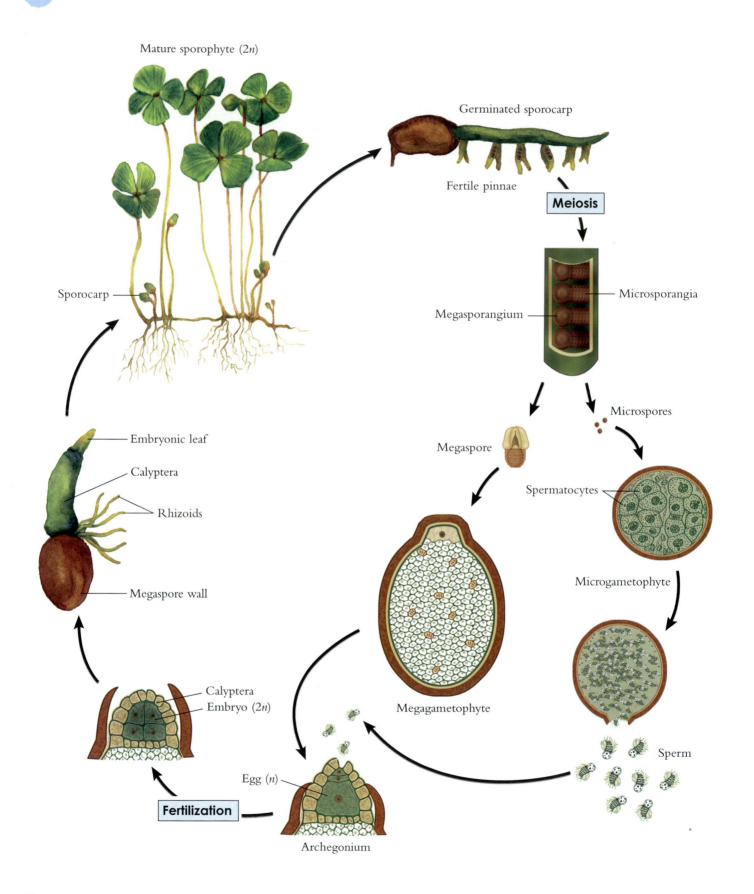

Mature sporophyte (2n)

Germinated sporocarp

Fertile pinnae

Meiosis

Sporocarp

Microsporangia

Megasporangium

Microspores

Embryonic leaf

Calyptera

Rhizoids

Megaspore

Spermatocytes

Megaspore wall

Microgametophyte

Calyptera
Embryo (2n)

Megagametophyte

Sperm

Egg (n)

Fertilization

Archegonium

Figure 7.66 Life cycle of the heterosporous water fern, *Marsilea* sp.

Figure 8.53 First-year ovulate cone in *Pinus* sp.
1. Pollen cones   2. First-year ovulate cone

Figure 8.54 Longitudinal section through a first-year ovulate cone in *Pseudotsuga* sp.
1. Cone scale bracts   2. Immature ovule

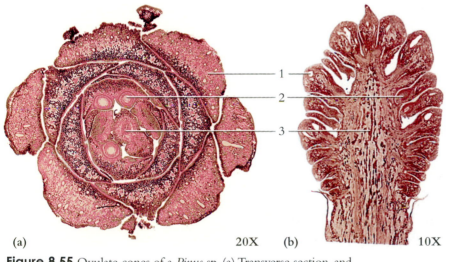

(a)   20X   (b)   10X

Figure 8.55 Ovulate cones of a *Pinus* sp. (a) Transverse section, and (b) longitudinal section.
1. Ovuliferous scale   2. Ovule   3. Cone axis

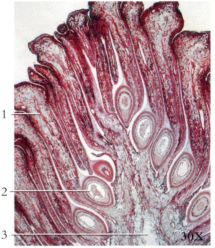

30X

Figure 8.56 Magnified view of a *Pinus* sp. ovulate cone (longitudinal view).
1. Ovuliferous scale   3. Cone axis
2. Ovule

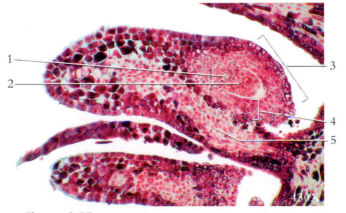

Figure 8.57 Magnified view of a *Pinus* sp. ovule (immature).
1. Megaspore mother cell   4. Integument
2. Nucellus   5. Cone scale
3. Ovule

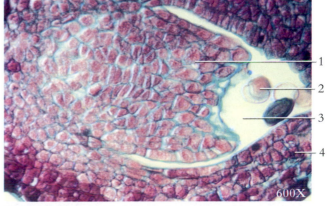

600X

Figure 8.58 Magnified view of an ovule of *Pinus* sp. with pollen grains in the pollen chamber.
1. Nucellus   3. Pollen chamber
2. Pollen grain   4. Integument

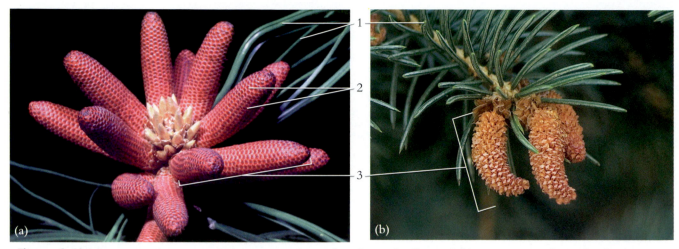

Figure 8.59 Microsporangiate cones of (a) *Pinus* sp. prior to the release of pollen and (b) *Picea pungens* after pollen has been released. The pollen cones are at the end of a branch.

1. Needle-like leaves   2. Microsporophylls   3. Pollen cone

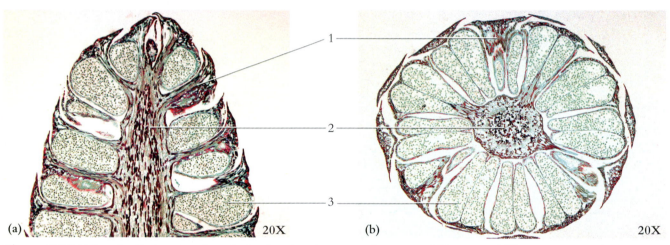

Figure 8.60 (a) Longitudinal section through the tip of a microsporangiate cone of *Pinus* sp. and (b) a transverse section.

1. Sporophyll   2. Cone axis   3. Microsporangium

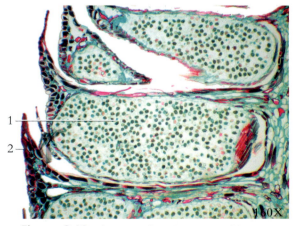

Figure 8.61 Close up of a microsporangiate cone scale and microsporangium of *Pinus* sp.

1. Microsporangium   2. Microsporophyll
   with pollen grains

Figure 8.62 Micrograph of stained pollen grains of *Pinus* sp. showing wings.

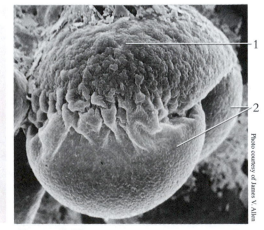

Figure 8.63 Scanning electron micrograph of a *Pinus* sp. pollen grain with inflated bladder-like wings.

1. Pollen body   2. Wings

Photo courtesy of James V. Allen

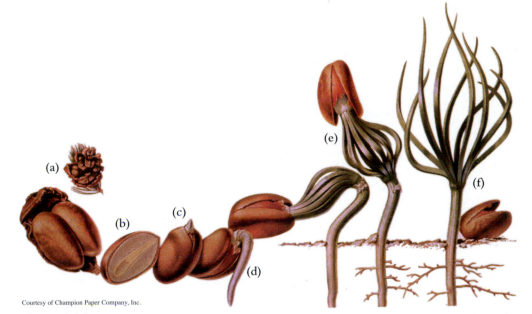

Courtesy of Champion Paper Company, Inc.

Figure 8.64 Diagram of pinyon pine seed germination producing a young sporophyte. (a) The seeds are protected inside the cone, two seeds formed on each scale. (b) A sectioned seed showing an embryo embedded in the female gametophyte tissue. (c) The growing embryo splits the shell of the seed, enabling the root to grow toward the soil. (d) As soon as the tiny root tip penetrates and anchors into the soil, water, and nutrients are absorbed. (e) The cotyledons emerge from the seed coat and create a supply of chlorophyll. Now the sporophyte can manufacture its own food from water and nutrients in the soil and carbon dioxide in the air. (f) Growth occurs at the terminal buds at the base of the leaves.

Figure 8.65 Young sporophyte (seedling) of a pine, *Pinus* sp.
1. Seedling leaves (needles)
2. Young stem
3. Young roots

Figure 8.66 Close up of an ovulate cone scale in *Pinus* sp.
1. Matureseeds (wings)
2. Ovulatecone scale
3. Seed(containing embryo within seed coat)

Figure 8.67 Young ovule of *Pinus* sp. showing the megagametophyte.
1. Ovule
2. Micropyle
3. Archegonium
4. Megagametophyte

Figure 8.68 Young ovule of *Pinus* sp. showing the egg in archegonium.
1. Egg
2. Nucleus

Figure 8.69 Magnified view of the ovule of *Pinus* sp. showing the embryo.
1. Integument
2. Leaf primordium
3. Root primordium

Figure 8.70 Monkey puzzle pine, *Araucaria auracana*, is a primitive conifer characterized by sharp, thick spine-like leaves. (a) Tree, (b) stems and (c) trunk.

Figure 8.71 Leaves of most species of conifers are needle-shaped such as those of the blue spruce, *Picea pungens* (a). *Araucaria heterophylla,* Norfolk Island pine, however, (b) has awl-shaped leaves, and *Podocarpus* sp. (c) has strap-shaped leaves.

Figure 8.72 Conifer fossils from the Mesozoic Era. (a) Two small branches from the genus *Taxodium*; and (b) a petrified cone of *Pinus* sp.

Figure 8.73 *Ephedra* sp. shrubs in Capitol Reef National Park.

Figure 8.74 *Ephedra* sp. is one of three genera of shrubs within the phylum Gnetophyta. Although found throughout most arid or semiarid regions of the world, *Ephedra* sp. is the only one of the three genera of gnetophytes found in the United States. It is a highly branched shrub with very small leaves.

Figure 8.75 Mormon tea, *Ephedra* sp., is a small shrub within the family Ephedraceae. Its common name comes from its use by pioneers in the American West to make a hot beverage.

Figure 8.76 Stems and scale-like leaf of *Ephedra* sp.
1. Leaf 　　　　　　　　2. Stem

Figure 8.77 Microsporangiate cones of *Ephedra* sp.
1. Microsporangiate cones

10X

Figure 8.78 Stem of *Ephedra* sp. with several microsporangiate cones attached.
1. Stem 　　　　　　　2. Microsporangiate cone

Figure 8.79 *Ephedra* sp. with attached megasporangiate (ovulate) cones about the time of pollination.
1. Cones   2. Stem

Figure 8.80 *Ephedra* sp. with attached megasporangiate (ovulate) cones at time of seed release.
1. Cones   2. Stem

Figure 8.81 Stem of *Ephedra* sp. with four attached megasporangiate cones.
1. Cones   2. Stem

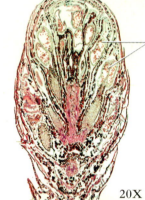

Figure 8.82 Longitudinal section through a microsporangiate cone of *Ephedra* sp.
1. Microsporangia

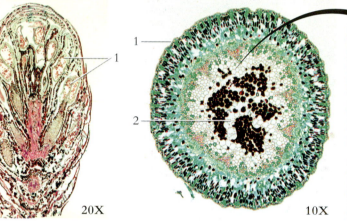

Figure 8.83 Transverse section through a stem of *Ephedra* sp. Note that unlike most other gymnosperms, *Ephedra* has vessel elements in the xylem similar to those found in angiosperms.

1. Epidermis   3. Cuticle   5. Phloem
2. Pith   4. Cortex   6. Xylem

Figure 8.84 Young plant of *Welwitschia mirabilis*. a gnetophyte that grows in the coastal desert of southwestern Africa. Most of this unusual plant is buried in sandy soil. The exposed portion consists of a woody disk that produces two strap-shaped leaves.

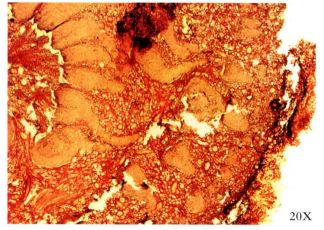

Figure 8.85 Transverse section through a young stem of *Welwitschia mirabilis*. Cone-bearing branches arise from meristematic tissue on the margin of the disk.

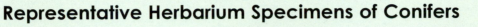

Representative Herbarium Specimens of Conifers

Figure 8.86 Herbarium specimen of *Phyllocladus aspleniifolius*. Found in New Zealand and Australia, *Phyllocladus* is a primitive conifer. Though *Phyllocladus* superficially resembles some angiosperms, it is actually a cone-bearing species.

Figure 8.87 Herbarium specimen of *Thuja orientalis*. A cultivated tree of Old World origin, this tree provides seeds for many species of birds. *Thuja orientalis* is a member of family Cupressaceae.

Figure 8.88 Herbarium specimen of *Pinus flexilis*. This species is a high altitude tree within the family Pinaceae.

Figure 8.89 Herbarium specimen of the giant sequoia (or Sierra redwood), *Sequoiadendron giganteum*. The giant sequoia is within the family Taxodiaceae. Once widespread throughout temperate climates of North America, the giant sequoia occurs now in isolated groves in the Sierra Nevada mountain range of Northern California. Many trees of the Sequoia National Park are over two thousand years old.

Figure 8.90 Herbarium specimen of a juniper, *Juniperus*. Within the family Cupressaceae, this tree along with the pinyon pine, comprises the dominant vegetation type within the pinyon-juniper forest in semiarid environments of the Western United States. Juniper berries are used to flavor gin.

Figure 8.91 Herbarium specimen of a larch, *Larix dahurica*, family Pinaceae. This larch occurs in high latitude near the Arctic Circle. The leaves of the larch are deciduous and spirally arranged.

Conifers of Commercial Value as Wood Products

Lodgepole pine–*Pinus contorta*

Cedar
Western red cedar (Western arborvitae)–*Thuja plicata*.
Distribution western North America. Size 150 to 200 feet in height. Commercially important for home construction. Excellent wood for shingles, house siding, frames and doors, cabinets, caskets, boat trimming, and fence posts.

Fir
Douglas fir–*Pseudotsuga menziesii*.
Distribution in the mountainous areas of western North America, comprises more than 50% of many western forest species. Size 150 to 200 feet in height. Commercially important for construction of homes (framing and plywood), boat building, railroad ties and mining timber, pulp for paper, pallets and crates, and wooden boxes.

Hemlock
Eastern hemlock–*Tsuga canadensis*.
Distribution eastern North America. Size 80 to 100 feet in height. Commercially important for construction of homes–doors and door frames, cabinets, and boxes; tannins used for tanning animal hides for leather.

Redwood
California redwood–*Sequoia sempervirens*.
Distribution California and parts of northwestern United States. Size 300 to 330 feet in height (the tallest species of tree). Commercially important for construction of decks and outdoor furniture, shingles, caskets, and supporting lumber for homes.

Pine
Loblolly pine–*Pinus taeda*.
Distribution eastern United States. Size 60 to 90 feet in height; grows in a wide variety of soils, but thrives in wet, acidic soil. One of the leading commercial timber species in the United States; used in construction of homes—interior wood finishing, flooring and wainscoting.

Ponderosa pine–*Pinus ponderosa*.
Distribution mountainous regions of western United States. Size 150 to 225 feet in height. Intolerant of shade; resistant to drought; tolerates alkaline soils. Commercially important for construction of caskets, fences, railroad ties, heavy timber, and log homes.

Slash pine–*Pinus elliottii*.
Distribution United States Size 60 to 90 feet in height. Commercially important for pulpwood and construction of heavy timber, railroad ties and veneer; baskets, boxes, and crates.

Lodgepole pine–*Pinus contorta*
Distribution northern Rocky mountains and parts of northwestern United States. Size 60 to 100 feet in height. Commercially important for construction of utility poles and log homes.

Sugar pine–*Pinus lambertiana*.
Distribution Cascade mountain range in western United States. Size 180 to 220 feet in height. Commercially important for construction of pallets, boxes and crates, doors and cabinets, organ pipes and piano keys.

Western white pine– *Pinus monticola*.
Distribution western North America. Size 150 to 170 feet in height. Commercially important for construction of homes and home products–plywood, door frames, tabletops; matches and boxes.

Spruce
Red spruce–*Picea rubens*.
Distribution eastern North America. Size 60 to 75 feet in height. Commercially important for pulp for paper, musical sounding boards, ladders, crates; and boat building and oars.

Photos of Live Specimens of Conifers

Figure 8.92 Bald cypress (*Taxodium distichum*). Native to southeastern United States, the bald cypress grows in wet areas along rivers or in swamps. The height of a mature tree frequently ranges from 100 to 130 feet. The flattened and spiraled leaves are deciduous and become copper–colored in autumn. Prominent buttresses, called pneumatophores (c), frequently grow from the roots of a tree. Pneumatophores are adapted to utilize atmospheric oxygen. The cone (d) is rounded and changes color from green to brown upon ripening.

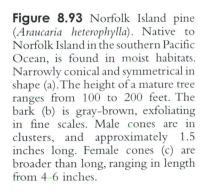

Figure 8.93 Norfolk Island pine (*Araucaria heterophylla*). Native to Norfolk Island in the southern Pacific Ocean, is found in moist habitats. Narrowly conical and symmetrical in shape (a). The height of a mature tree ranges from 100 to 200 feet. The bark (b) is gray-brown, exfoliating in fine scales. Male cones are in clusters, and approximately 1.5 inches long. Female cones (c) are broader than long, ranging in length from 4–6 inches.

Figure 8.94 Colorado blue spruce (*Picea pungens*). Native to Rocky Mountains in the United States on dry slopes and stream banks. The height of a mature tree ranges from 100 to 150 feet. It is a hardy medium-growing tree, narrowly conical in shape (a). Bark is gray-brown (b), and branches droop. Trees are bluish-grey in coloration (c).

Figure 8.95 Giant sequoia (*Sequoiadendron giganteum*). Native to California on moist west facing mountain slopes. Narrowly conical in shape (a) Bark is reddish-brown and fibrous (b). The height of a mature tree ranges from 260 to 300 feet. The leaves are awl-shaped and are 1 to 2.5 inches long. They are arranged spirally on the shoots (c).

Figure 8.96 Limber pine (*Pinus flexilis*). Native to Rocky Mountains of Western Northern America, limber pine is adapted to cold, harsh winters. It obtains heights of 30 to 50 feet and 15 to 35 feet in spread. This specimen was damaged by lightning (b) and is atypical in form. The needles are 2 to 4 inches long and are arranged in fascicles of 5 (c).

Figure 8.97 Monterey pine (*Pinus radiata*) is native to the western United States. Found in dry coastal habitats, it is broadly conical, becoming rounded to flat in shape. The bark is reddish brown to gray and furrowed (b). The needles are arranged 2 or 3 per fascicle. Seed cones (c) require 2 years to mature. The height of a mature tree ranges from 85 to 100 feet.

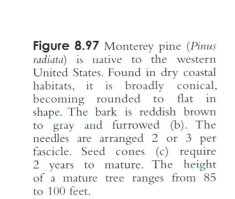

Chapter 9
Angiospems: Enclosed Seed Plants Flowering Plants

Angiosperms are plants that produce flowers and fruits. They range in size from the minute duckweed (1 mm in total size) to *Eucalyptus* trees (100 m tall). A few are saprophytic or parasitic, but most are free-living. Angiosperms, because of their rapid life cycles, proliferate in all major habitats, including terrestrial, marsh, fresh water, and marine.

Magnoliophyta (=Anthophyta) is the most recent plant phylum to have evolved. Flowering plants have only existed since the late Jurassic or early Cretaceous, perhaps 130 million years ago, but currently dominate the Earth's terrestrial flora in terms of variety and biomass. Most flowering plants produce a crop of mature seeds in a single year, and many species are able to pass from a seed to a mature seed-producing plant in a matter of a few weeks. This rapid cycle has allowed these plants to be efficient in occupying new territory. Many gymnosperms, require two or more years to complete their life cycle.

Angiosperms are divided into the class *Monocotyledonae* (monocots), including the grasses, palms, lilies, and orchids; and class *Dicotyledonae* (dicots), including most other familiar flowers, shrubs, and trees (except conifers). An estimated 65,000 species of monocots and 170,000 species of dicots have been identified, although many botanists believe the actual number of species will turn out to be much higher.

Like most plants, angiosperms consist of a vegetative portion and a germinative portion. The vegetative portion includes the roots, stems, and leaves, which are involved primarily with the manufacture and transfer of food and plant growth. The germinative portion includes the flowers and is involved primarily with sexual reproduction and the formation of seeds.

Table 9.1 Angiosperms within the Kingdom Plantae

Classes and Representative Kinds	Characteristics
Class Monocotyledonae— monocots	One cotyledon; leaf veins usually parallel; vascular bundles within stem scattered; fibrous root system; floral parts usually in multiples of three
Class Dicotyledonae— dicots	Two cotyledons; leaf veins usually net-like; vascular bundles within stem arranged in ring; taproot usually present; floral parts usually in multiples of four or five

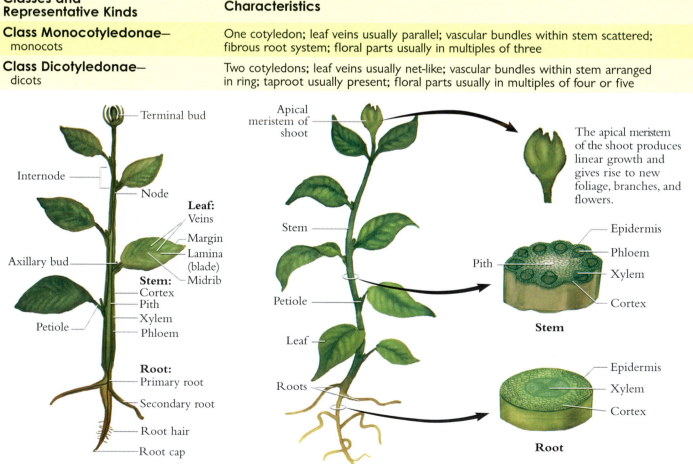

Figure 9.1 A diagram illustrating the anatomy of a typical dicot.

Figure 9.2 A diagram illustrating the principal organs and tissues of a typical dicot.

Phylum Magnoliophyta (=Anthophyta) – angiosperms: monocots and dicots

Figure 9.3 Comparison and examples of monocots and dicots.

Roots of Angiosperms

Plants are only as healthy as are their root systems. The root system of an angiosperm is the descending, usually underground portion of the plant. The roots can make up more than half the plant body. Roots anchor the plant. Water and nutrients are absorbed, stored, and conducted by the roots. The root system of a plant is influenced by soil type and mineral content and the amount and timing of moisture.

Monocots, such as grasses, generally have fibrous, or diffuse, root systems. Dicots, such as shrubs and most woody plants, generally have taproot systems. Specialized supporting root systems include aerial roots and prop roots. Taproots, such as found in carrots and turnips, are capable of storing large amounts of food.

The active root system of most angiosperms consists of four main regions:

Root cap—A root cap region is a cluster of cells at the tip of the root. It protects the root during growth through the soil.

Meristematic region—The meristematic region of the root, covered by the root cap, is where new cells are added to the growing root by active cell division.

Elongation region—The elongation region of the root is where newly added cells increase in size.

Maturation region—The maturation region of the root is where cells differentiate into epidermal and cortex layers and vascular tissues (xylem and phloem). The root hairs formed in the epidermis of this region greatly increase the surface area for absorption. The cortex stores reserve food.

Fibrous root system
(grasses)

Taproot
(shrubs)

Modified taproot
(carrot)

Prop roots
(corn)

Aerial roots
(orchid)

Figure 9.4 Root systems of angiosperms.

(a)

(b)

Figure 9.5 Root system of an orchid (monocot) (a) showing aerial roots and a banyan (dicot) (b) showing prop roots. Monocot roots are fibrous, with many roots of more or less equal size. Dicots usually have a taproot system, consisting of a long central root with smaller, secondary roots branching from it.

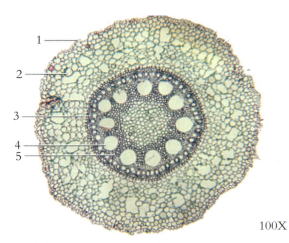

Figure 9.6 Transverse section of the root of the monocot *Smilax* sp.
1. Epidermis   4. Xylem
2. Cortex   5. Phloem
3. Endodermis

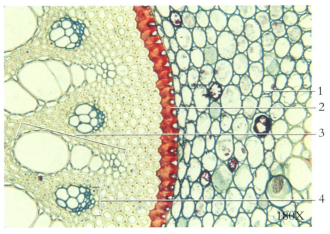

Figure 9.7 Close up of a root of the monocot *Smilax* sp.
1. Cortex   3. Xylem
2. Endodermis   4. Phloem

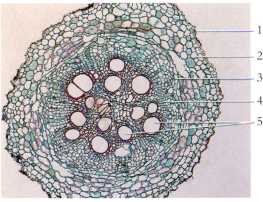

Figure 9.8 Transverse section of a sweet potato root, *Ipomaea* sp.
1. Remnants of epidermis   4. Phloem
2. Cortex   5. Xylem
3. Endodermis

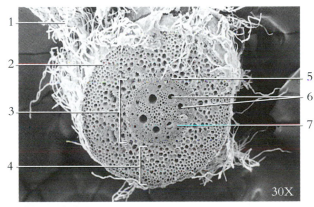

Figure 9.9 Scanning electron micrograph of a young root of wheat, *Triticum* sp., showing root hairs.
1. Root hair   5. Endodermis
2. Epidermis   6. Primary xylem
3. Stele   7. Primary phloem
4. Cortex

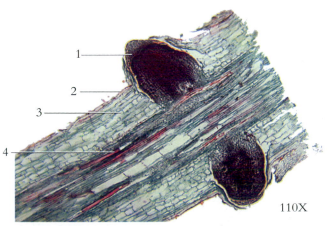

Figure 9.10 Longitudinal section of a willow species showing lateral root formation.
1. Lateral root   3. Cortex
2. Epidermis   4. Vascular tissue

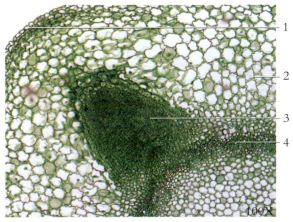

Figure 9.11 Transverse section branch root formation of *Phaseolus* sp.
1. Epidermis   3. Branch root
2. Cortex   4. Vascular tissue (stele)

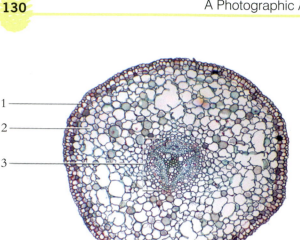

100X

Figure 9.12 Transverse section of a young root of *Salix* sp.

1. Epidermis   3. Stele
2. Cortex

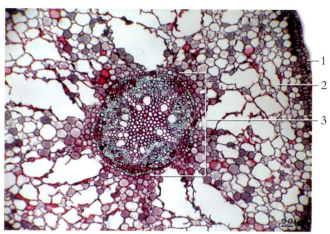

220

Figure 9.13 Transverse section of an older root of *Salix* sp., showing early secondary growth.

1. Epidermis   3. Vascular tissue
2. Cortex

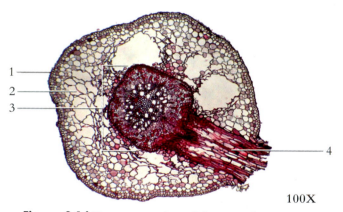

100X

Figure 9.14 Transverse section of the root of *Salix* sp. showing branch root development.

1. Epidermis   3. Stele
2. Cortex   4. Branch root

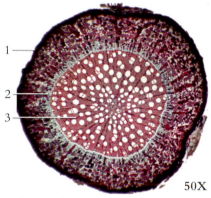

50X

Figure 9.15 Transverse section of an older root of *Salix* sp., showing secondary growth.

1. Periderm   3. Secondary
2. Secondary   xylem
     phloem

100X

Figure 9.16 Transverse section of a young root of *Pyrus* sp.

1. Epidermis   4. Primary xylem
2. Cortex   5. Endodermis
3. Primary phloem

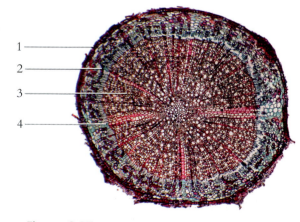
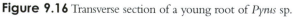

50X

Figure 9.17 Transverse section of the root, *Pyrus* sp., showing secondary growth.

1. Periderm   3. Secondary xylem
2. Secondary phloem   4. Vascular cambium

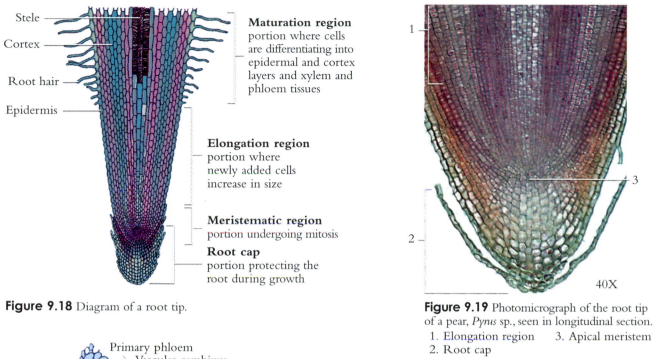

Stele

Cortex

Root hair

Epidermis

Maturation region
portion where cells are differentiating into epidermal and cortex layers and xylem and phloem tissues

Elongation region
portion where newly added cells increase in size

Meristematic region
portion undergoing mitosis

Root cap
portion protecting the root during growth

Figure 9.18 Diagram of a root tip.

40X

Figure 9.19 Photomicrograph of the root tip of a pear, *Pyrus* sp., seen in longitudinal section.
1. Elongation region   3. Apical meristem
2. Root cap

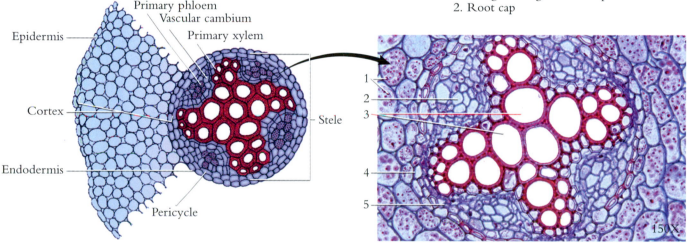

Epidermis

Primary phloem
Vascular cambium
Primary xylem

Cortex

Endodermis

Stele

Pericycle

150X

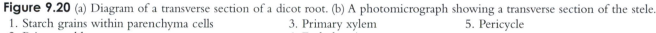

Figure 9.20 (a) Diagram of a transverse section of a dicot root. (b) A photomicrograph showing a transverse section of the stele.
1. Starch grains within parenchyma cells   3. Primary xylem   5. Pericycle
2. Primary phloem   4. Endodermis

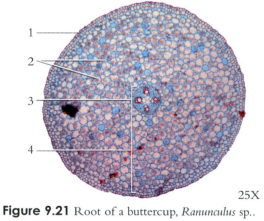

25X

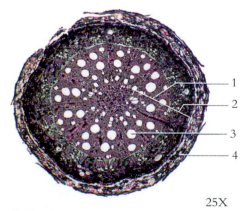

25X

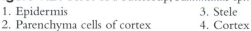

Figure 9.21 Root of a buttercup, *Ranunculus* sp..
1. Epidermis   3. Stele
2. Parenchyma cells of cortex   4. Cortex

Figure 9.22 Transverse section of the root of basswood, *Tilia* sp., showing secondary growth.
1. Secondary xylem   3. Vessel element
2. Secondary phloem   4. Periderm

Stems of Angiosperms

The stem of an angiosperm is usually the ascending portion of the plant. It produces and supports leaves and flowers; transports and stores water and nutrients; and provides growth through cell division. Branches and twigs, smaller extensions and/or branches of the stem of certain plants, directly support leaves and flowers.

Herbaceous stems are often soft and succulent. The herbaceous stems of monocots have scattered vascular bundles, while herbaceous stems of dicots have vascular bundles arranged in a ring. Some species of monocots have stems reinforced with secondary fibrous strands and appear woody.

Woody stems are hardened and often larger in diameter. They usually increase in diameter by secondary growth originating from a lateral *cork cambium* and a *vascular cambium*.

Woody stems of dicots often have three parts:
1. *Bark*, which contains *periderm* and phloem;
2. *Wood,* which contains the annual rings of secondary xylem;
3. *Pith*, composed of loosely packed parenchyma cells at the center of the vascular tissue.

Linear growth of woody dicot stems occurs at *terminal buds,* where mitosis occurs at the *apical meristem*. Buds also contain developing leaves and, in certain locations, developing flowers. *Nodes* of branches or twigs are the points of leaf attachment, and *internodes* are the spaces between the nodes. *Lenticels* are pores in the bark which facilitate gas exchange.

People use stems for products including paper, building materials, furniture, and fuel. In addition, food is obtained from the stems of potatoes, onions, asparagus, and other plants.

Figure 9.23 Examples of the variety and specialization of angiosperm stems. The stem of an angiosperm is often the ascending portion of the plant specialized to produce and support leaves and flowers, transport and store water and nutrients, and provide growth through cell division. Stems of plants are utilized extensively by humans in products including paper, building materials, furniture, and fuel. In addition, the stems of potatoes, onions, cabbage, and other plants are important food crops.

Figure 9.24 Specialized underground stems.
(a) A potato (tuber) and (b) an onion (bulb).
1. Node (eye) bearing a minute scale leaf and stem bud
2. Bulb scales (modified leaves)
3. Short stem

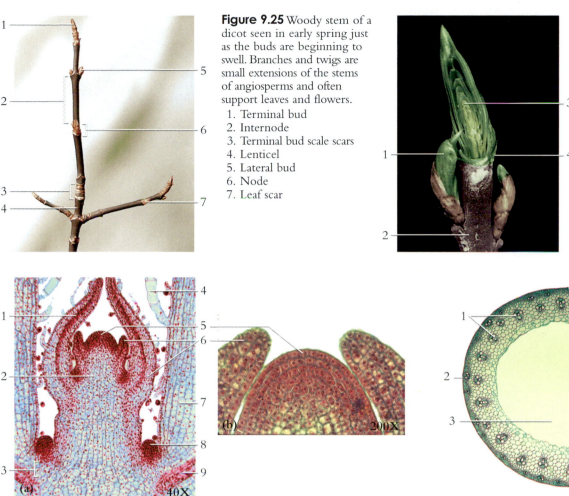

Figure 9.25 Woody stem of a dicot seen in early spring just as the buds are beginning to swell. Branches and twigs are small extensions of the stems of angiosperms and often support leaves and flowers.
1. Terminal bud
2. Internode
3. Terminal bud scale scars
4. Lenticel
5. Lateral bud
6. Node
7. Leaf scar

Figure 9.26 Terminal bud of a woody stem that has been longitudinally sectioned to show developing leaves.
1. Lateral bud
2. Stem
3. Leaf primordia
4. Bud scale

Figure 9.27 (a) Longitudinal section of the stem tip of the common houseplant *Coleus* sp. (b) Magnified view of apical meristem.
1. Procambium
2. Ground meristem
3. Leaf gap
4. Trichome
5. Apical meristem
6. Developing leaf primordia
7. Leaf primordium
8. Axillary bud
9. Developing vascular tissue

Figure 9.28 Transverse section through the stem of a monocot, *Triticum* sp., wheat.
1. Vascular bundles
2. Epidermis
3. Ground tissue cavity
4. Parenchyma cells

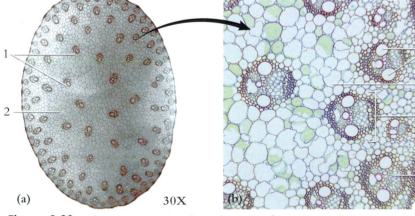

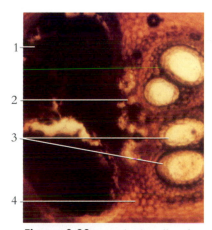

Figure 9.29 (a) Transverse section from the stem of a monocot, *Zea mays* (corn). (b) Magnified view.
1. Vascular bundles with primary xylem and phloem
2. The pattern of vascular bundles in a monocot is known as an atactostele
3. Epidermis
4. Vessel elements of primary xylem
5. Parenchyma cells
6. Vascular bundle
7. Primary phloem

Figure 9.30 Vascular bundle of a fossil palm plant.
1. Bundle cap (fibers)
2. Phloem
3. Vessel elements
4. Ground tissue (parenchyma)

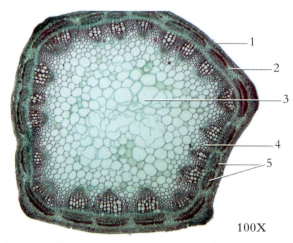

Figure 9.31 Transverse section through a dicot stem, *Trifolium* sp., clover, showing an eustele.

1. Epidermis
2. Cortex
3. Pith
4. Interfascicular region
5. Vascular bundles with caps of phloem fibers

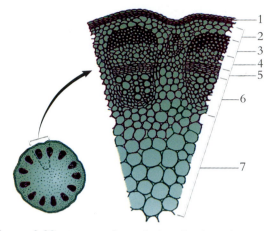

Figure 9.32 Diagram of vascular bundles from the stem of a dicot showing the eustele.

1. Early periderm
2. Cortex
3. Phloem fibers
4. Phloem
5. Vascular cambium
6. Xylem
7. Pith

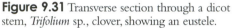
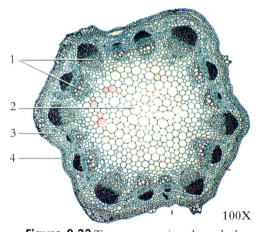

Figure 9.33 Transverse section through the stem of a young sunflower, *Helianthus* sp.

1. Vascular bundles
2. Pith
3. Cortex
4. Epidermis

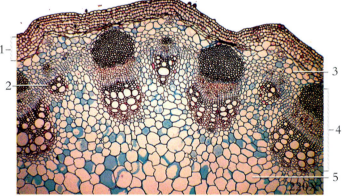

Figure 9.34 Transverse section through an older stem of a sunflower, *Helianthus* sp.

1. Collenchyma
2. "New" vascular bundle
3. Cortex
4. "Original" vascular bundle with secondary growth
5. Pith

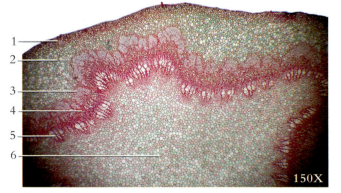

Figure 9.35 Young oak stem showing early secondary growth.

1. Epidermis and developing periderm
2. Cortex
3. Phloem
4. Vascular cambium
5. Xylem
6. Pith

Figure 9.36 Longitudinal section through the secondary xylem and phloem of *Juglans* sp. (walnut).

1. Secondary phloem
2. Ray
3. Secondary xylem
4. Vessel element
5. Fibers

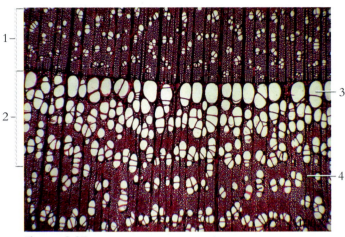

Figure 9.37 Transverse section through a stem of *Cercis canadensis*, showing the pattern of fibers and vessel elements.
1. Late wood
2. Early (spring) wood
3. Vessel element
4. Fibers

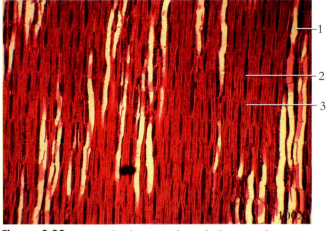

Figure 9.38 Longitudinal section through the secondary xylem (wood) of *Cercis canadensis*.
1. Vessel elements
2. Ray
3. Fibers

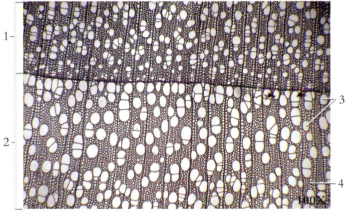

Figure 9.39 Transverse section through the secondary xylem (wood) of the stem of *Populus* sp., showing the less seasonal distribution of vessel elements.
1. Late wood
2. Early (spring) wood
3. Vessel elements
4. Fibers

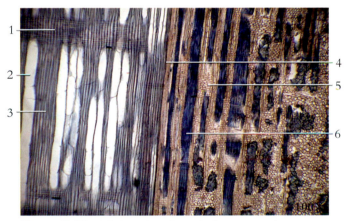

Figure 9.40 Radial section through the stem at the cambium of *Populus* sp., showing both phloem and xylem.
1. Ray in xylem
2. Vessel element
3. Xylem fibers
4. Cambium
5. Ray in phloem
6. Phloem fibers

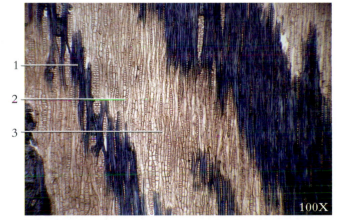

Figure 9.41 Longitudinal section through the secondary phloem of the stem of *Populus* sp., showing the distribution of the fibers and phloem tissue (sieve tube elements).
1. Phloem fibers
2. Phloem ray
3. Sieve tube elements

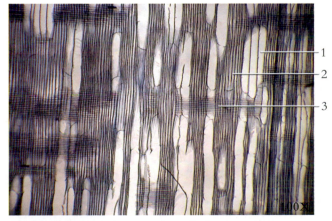

Figure 9.42 Radial section through the secondary xylem (wood) of the stem of *Populus* sp., showing the distribution pattern of the fibers and vessel elements.
1. Vessel element
2. Xylem fiber
3. Xylem ray

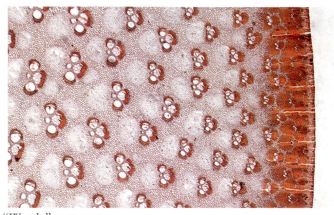

"Woody" monocot

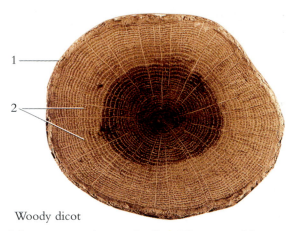

Woody dicot

Figure 9.43 Comparison of the transverse sections of stems of a "woody" monocot and a woody dicot. The stem of the "woody" monocot is rigid because of the fibrous nature of the numerous vascular bundles. The stem of the woody dicot is rigid because of the compact xylem cells impregnated with lignin forming the dense, hardened wood, seen as annual rings.

1. Bark   2. Annual rings

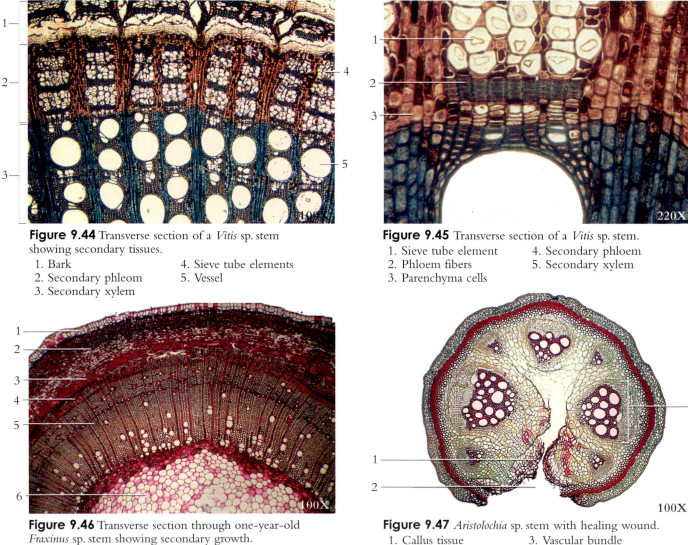

Figure 9.44 Transverse section of a *Vitis* sp. stem showing secondary tissues.

1. Bark   4. Sieve tube elements
2. Secondary phloem   5. Vessel
3. Secondary xylem

Figure 9.45 Transverse section of a *Vitis* sp. stem.

1. Sieve tube element   4. Secondary phloem
2. Phloem fibers   5. Secondary xylem
3. Parenchyma cells

Figure 9.46 Transverse section through one-year-old *Fraxinus* sp. stem showing secondary growth.

1. Periderm   4. Secondary phloem
2. Cortex   5. Secondary xylem
3. Phloem fibers   6. Pith

Figure 9.47 *Aristolochia* sp. stem with healing wound.

1. Callus tissue   3. Vascular bundle
2. Wound

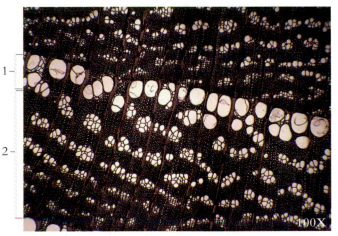

Figure 9.48 Transverse section through secondary xylem (wood) of *Ulmus* sp. showing a ring porous arrangement of the vessel elements.

1. Early (spring) wood with large vessel elements
2. Late (spring) wood with smaller vessel elements

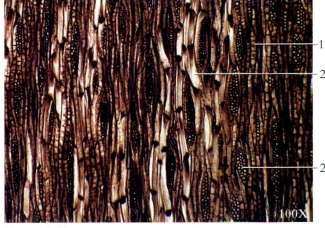

Figure 9.49 Tangential section through the secondary phloem of *Ulmus* sp.

1. Parenchyma
2. Sieve tube element
3. Ray

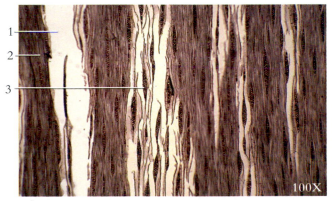

Figure 9.50 Tangential section through the secondary xylem (wood) of *Ulmus* sp.

1. Vessel elements
2. Xylem fibers
3. Ray

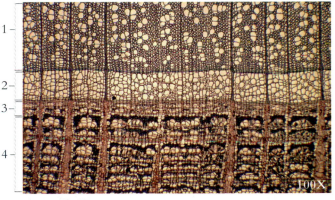

Figure 9.51 Transverse section through the secondary xylem (wood) and secondary phloem of *Tilia* sp. Note the diffuse porous vessel elements in the wood.

1. Secondary xylem from the previous year
2. Secondary xylem from the present year
3. Secondary phloem from the present year
4. Secondary phloem from the previous year(s)

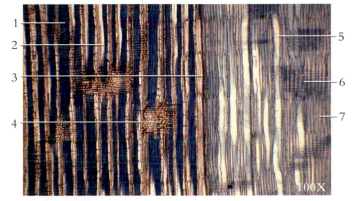

Figure 9.52 Radial section through the secondary xylem phloem of *Tilia* sp.

1. Phloem fibers
2. Sieve tube elements
3. Vascular cambium
4. Phloem ray
5. Vessel element
6. Xylem ray
7. Xylem fibers

Figure 9.53 Longitudinal section through the secondary xylem (wood) of *Tilia* sp.

1. Vessel element
2. Xylem ray
3. Xylem fibers

(a)   (b)   (c)   (d)   (e)   (f)

Figure 9.54 Samples of bark patterns of representative conifers and angiosperms

(a) **Redwood**—The tough, fibrous bark of a redwood tree may be 30 cm thick. It is highly resistant to fire and insect infestation.

(b) **Ponderosa pine**—The mosaic-like pattern of the bark of mature ponderosa pine is resistant to fire.

(c) **White birch**—The surface texture of bark on the white birch is like white paper. The bark of the white birch was used by Indians in Eastern United States for making canoes.

(d) **Sycamore**—The mottled color of the sycamore bark is due to a tendency for large, thin, brittle plates to peel off, revealing lighter areas beneath. These areas grow darker with exposure, until they, too, peel off.

(e) **Mangrove**—The leathery bark of a mangrove tree is adaptive to brackish water in tropical or semi-tropical regions.

(f) **Shagbark hickory**—The strips of bark in a mature shagbark hickory tree gives this tree its common name.

Figure 9.55 Angiosperm, *Ruscus aculeatus*, is characterized by stems (a) that resemble leaves in form and function. Note the true leaf (b) arising from the leaf-like stem.

1. Stem   2. Leaf   3. Flower bud

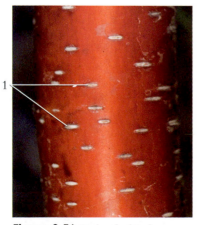

Figure 9.56 Bark of a birch tree, *Betula occidentalis*, showing lenticels. Lenticels are spongy areas in the cork surfaces that permit gas exchange between the internal tissues and the atmosphere.

1. Lenticels

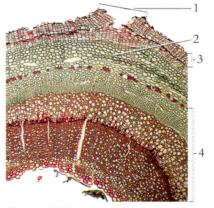

Figure 9.57 Transverse section of a dicot stem showing a lenticel and stem tissues.

1. Lenticel   3. Periderm
2. Cortex   4. Vascular tissue

Figure 9.58 Gall on an oak, *Quercus* sp., stem. The feeding of a gall wasp larva causes abnormal growth and the formation of a gall. The wasp larva feeds upon the gall tissue, pupates within this enclosure, and then chews an exit to emerge.

1. Gall   2. Stem

Leaves of Angiosperms

The leaf of an angiosperm manufactures food by *photosynthesis*, which is the production of sugar from carbon dioxide and water, in the presence of chlorophyll, with sunlight as the source of energy.

Leaves are attached at *nodes* of a stem by *petioles*. The leaf *midrib* is vascular tissue (a *vein*) continuous with the vascular tissue of the petiole through the leaf *lamina* (blade) and gives rise to numerous branching secondary veins. Leaves may be classified on the basis of arrangement on a petiole, the arrangement of the veins, and the appearance of the margins. A *deciduous* leaf is one that is shed during the autumn season as the petiole detaches from the stem.

The typical tissue arrangement of a leaf includes an *upper epidermis*, a *lower epidermis* and the centrally located *mesophyll*. The cells of the mesophyll contain *chloroplasts*, which are necessary for photosynthesis. Mesophyll is often divided into *palisade mesophyll* and *spongy mesophyll*. Veins within the mesophyll conduct material through the leaf. Atmospheric gases containing carbon dioxide enter the leaf through *stomata*, the shape and opening of which is regulated by *guard cells*.

Leaves comprise the foliage of plants. Leaves provide habitat and food sources for many animals, including humans. They also provide protective ground cover and are the portion of the plant most responsible for oxygen replenishment into the atmosphere.

Platanus sp.

Allophylus sp.

Acer sp.

Platanus sp.

Figure 9.59 Compression fossils of four angiosperms leaves, approximately 50 million years old.

Venation	**Margin**	**Complexity**	**Arrangement on Stem**

Pinnate

Entire

Palmately compound

Opposite

Parallel

Pinnately lobed

Simple

Alternate

Palmate

Serrate

Pinnately compound

Whorled

Figure 9.60 Several representative angiosperm leaf types. Leaves comprise the foliage of plants, which provide habitat and a food source for many animals including humans. Leaves also provide protective ground cover and are the portion of the plant most responsible for oxygen replenishment into the atmosphere.

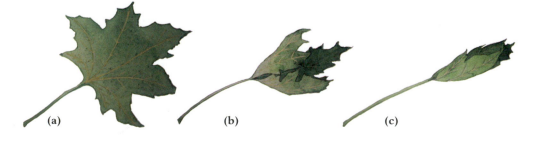

(a) (b) (c)

Figure 9.61 Shape of the leaf (a) is of adaptive value to withstand wind. As the speed of the wind increases (b) and (c), the leaf rolls into a tight cone shape, avoiding damage.

Figure 9.62 Angiosperm leaf showing characteristic surface features. Leaves are organs modified to carry out photosynthesis. Photosynthesis is the manufacture of food (sugar) from carbon dioxide and water, with sunlight providing energy.

1. Lamina (blade)   4. Veins
2. Serrate margin   5. Petiole
3. Midrib

Figure 9.63 Undersurface of an angiosperm leaf showing the vascular tissue lacing through the lamina, or blade, of the leaf.
1. Midrib
2. Secondary veins

Figure 9.64 Organic decomposition of a leaf is a gradual process beginning with the softer tissues of the lamina, leaving only the vascular tissues of the midrib and the veins, as seen in this photograph. With time, these will also decompose.

Figure 9.65 Brilliant autumn colors of leaves come about when colorless flavonoids are converted into anthocyanins as the chlorophyll breaks down.

Figure 9.66 Examples of specialized leaves for floatation. (a) Leaves from a giant water lily. (b) Water hyacinths, *Eichhornia* sp., have modified leaves that buoy the plants on the water surface. Water hyacinths are common in New World tropical fresh-water habitats, where they may become so thick that they choke out bottom-dwelling plants and clog waterways.

Figure 9.67 As seen on the leaflets in the upper right of this photograph, the leaves of the sensitive plant, *Mimosa pudica*, droop upon being touched. The drooping results from differential changes in turgor of the leaf cells in the pulvinus, a thickened area at the base of the leaflet.

Figure 9.68 Oleander, *Nerium oleander,* is a xerophyte (adapted to arid conditions), as reflected by rather thick, waxy leaves. Commonly, oleander plants in the American Southwest have brilliantly colored flowers. Oleander is native to Old World subtropics.

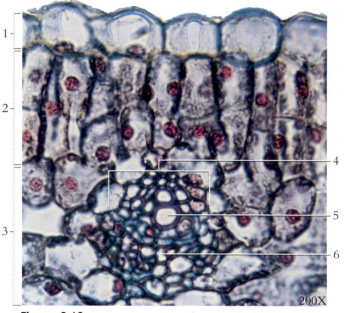

Figure 9.69 Transverse section of tomato leaf, *Lycopersicon* sp.

1. Upper epidermis	4. Leaf vein (vascular bundle)
2. Palisade mesophyll	5. Xylem
3. Spongy mesophyll	6. Phloem

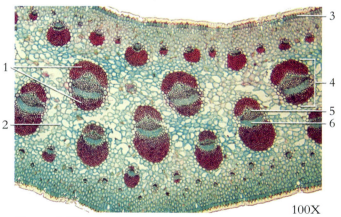

Figure 9.70 Transverse section of *Yucca* sp. leaf.

1. Bundle caps (fibers)	4. Vascular bundle
2. Ground tissue	5. Xylem
3. Epidermis	6. Phloem

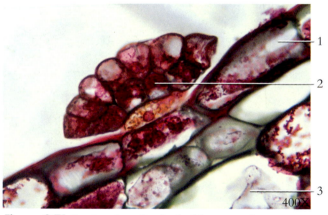

Figure 9.71 Section though a leaf of the venus flytrap, *Dionaea muscipula,* showing epidermal cells with an attached digestive gland. The gland is comprised of secretory parenchyma cells.

1. Epidermis	3. Mesophyll cell
2. Multicellular secretory gland	

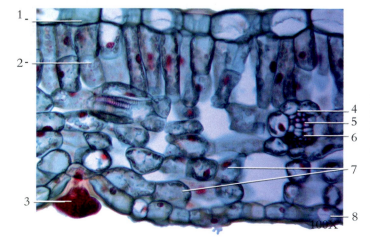

Figure 9.72 Transverse section through the leaf of the common hedge privet *Ligustrum* sp. The typical tissue arrangement of a leaf includes an upper epidermis, a lower epidermis, and the centrally located mesophyll. Containing chloroplasts, the cells of the mesophyll are often divided into palisade mesophyll and spongy mesophyll. Veins within the mesophyll conduct material through the leaf.

1. Upper epidermis	5. Xylem
2. Palisade mesophyll	6. Phloem
3. Gland	7. Spongy mesophyll
4. Bundle sheath	8. Lower epidermis

Figure 9.73 Longitudinal section of privet, *Ligustrum* sp., through the stem and petiole at the abscission layer.

1. Leaf petiole   1. Abscission layer

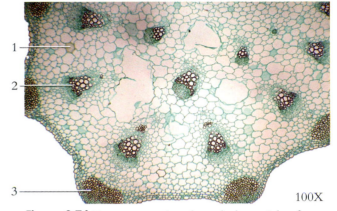

Figure 9.74 Transverse section through the petiole of wild carrot, *Daucus* sp., showing the leaf traces.

1. Ground tissue   3. Bundle of collenchyma
2. Leaf trace

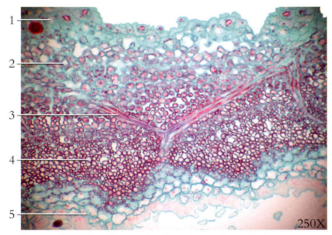

Figure 9.75 Paradermal leaf section of ivy, *Glechoma* sp., showing all leaf tissues.

1. Lower epidermis   4. Palisade mesophyll
2. Spongy mesophyll   5. Upper epidermis
3. Leaf vein

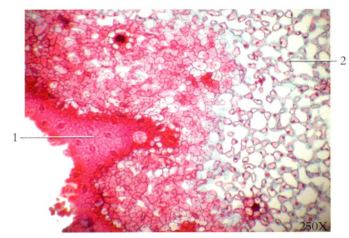

Figure 9.76 Paradermal leaf section of privet, *Ligustrum* sp., showing leaf tissues from lower epidermis through the spongy mesophyll.

1. Lower epidermis   2. Spongy mesophyll

Figure 9.77 Transverse section of bearberries, *Arctostaphylos* sp., through leaf.

1. Upper epidermis   4. Lower epidermis
2. Palisade mesophyll   5. Leaf vein
3. Spongy mesophyll

Figure 9.78 Transverse section of common garden flower, *Dianthus* sp., through leaf.

1. Upper epidermis   4. Lower palisade mesophyll
2. Upper palisade   5. Lower epidermis
   mesophyll   6. Leaf vein
3. Spongy mesophyll

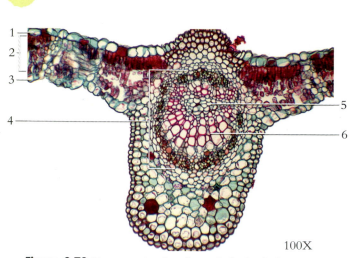

Figure 9.79 Transverse section through the leaf of basswood, *Tilia* sp.

1. Upper epidermis   4. Leaf vein (midrib)
2. Mesophyll   5. Phloem
3. Lower epidermis   6. Xylem

100X

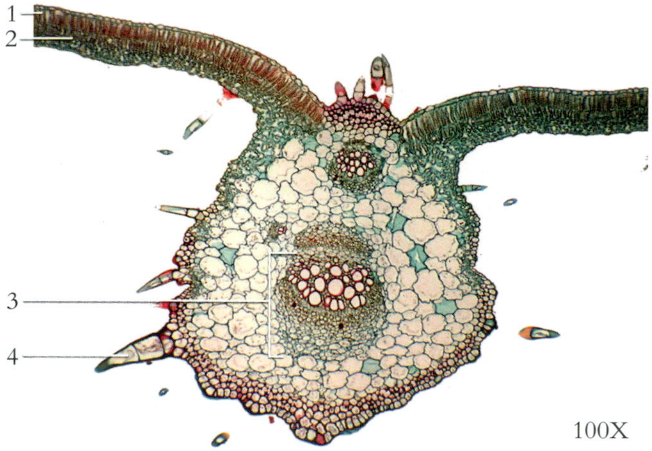

Figure 9.80 Transverse section through the leaf of cucumber, *Cucurbita* sp.

1. Palisade mesophyll   3. Leaf vein (midrib)
2. Spongy mesophyll   4. Trichome

100X

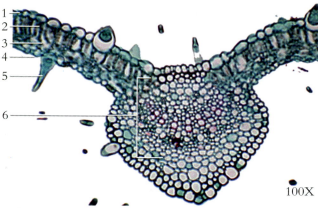

Figure 9.81 Transverse section through leaf of hemp, *Cannabis* sp.

1. Upper epidermis   4. Lower epidermis
2. Palisade mesophyll 5. Trichome
3. Spongy mesophyll   6. Leaf vein (midrib)

100X

Figure 9.82 Transverse section through the leaf of barberry, *Berberis* sp.

1. Upper epidermis   4. Lower epidermis
2. Palisade mesophyll 5. Leaf vein (midrib)
3. Spongy mesophyll

100X

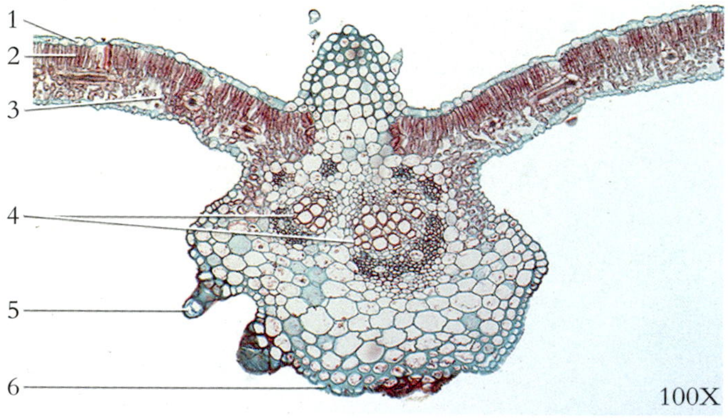

Figure 9.83 Transverse section through the leaf of sunflower, *Helianthus* sp.

1. Upper epidermis   4. Major leaf viens
2. Palisade mesophyll 5. Leaf hair
3. Spongy mesophyll   6. Lower epidermis

100X

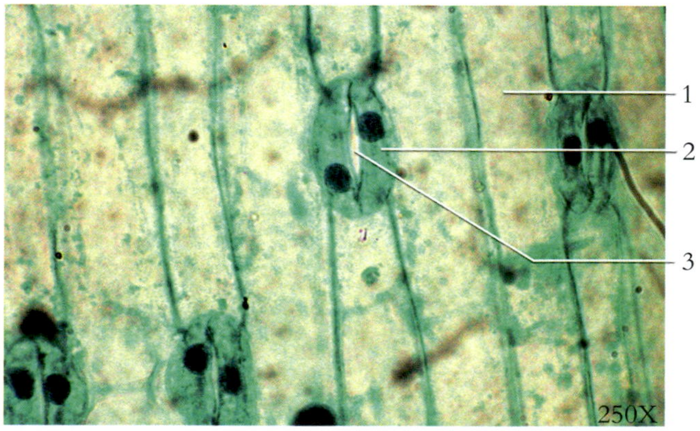

Figure 9.84 Face view of the epidermis of onion, *Allium* sp. Note the twin guard cells with the stoma opened.

1. Epidermal cell   3. Stomate
2. Guard cell

250X

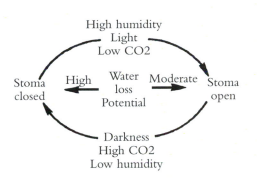

High humidity
Light
Low CO2

Stoma closed ← High | Water loss Potential | Moderate → Stoma open

Darkness
High CO2
Low humidity

Figure 9.85 Guard cells in many plants regulate the opening of the stomata according to the environmental factors, as indicated in this diagram. (a) Face view of a closed stoma of a geranium, and (b) an open stoma.

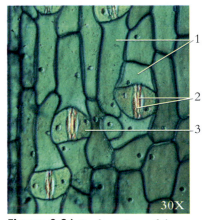

Figure 9.86 Surface view of the leaf epidermis of *Tradescantia* sp.
1. Epidermal cells
2. Guard cells surrounding stomata
3. Subsidiary cells

Figure 9.87 Leaves and epidermal hairs of a sundew, *Drosera capensis*.
1. Leaves
2. Epidermal hairs (glands)

Figure 9.88 Specialized leaves of the carnivorous pitcher plant, *Sarracenia* sp.

Figure 9.89 Leaves of the purple pitcher plant, *Sarracenia purpurea*, are adapted to entrap insects. The leaves are funnel-shaped and have epidermal hairs pointed toward the base of the leaf. Insects are attracted to the funnel where they are entrapped, die, and are digested by the plant.
1. Leaf   2. Epidermal hairs

Figure 9.90 Leaves of the venus flytrap, *Dionaea muscipula*, are adapted to entrap insects. An insect is attracted by nectar secreted on the surface of the leaf. The movement of the insect upon the leaves stimulates the sensitive trichomes on the upper surface of the leaves, triggering the leaves to close, entrapping the insect.

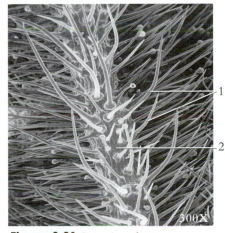

Figure 9.91 Scanning electron micrograph of a geranium leaf showing the prominent and abundant epidermal hairs.
1. Epidermal hairs
2. Epidermis

Figure 9.92 Joshua tree, *Yucca brevifolia*, is native to the Mojave desert. Its common name was derived from its resemblance to a bearded kneeling patriarch.

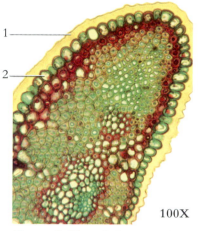

1

2

100X

Figure 9.93 Leaf of *Yucca* sp. shows a thick cuticle covering the epidermis of the leaf. The cuticle protects against excessive water loss.

   1. Cuticle   2. Epidermis

Figure 9.94 *Euphorbia* sp., a member of the spurge family, is specialized to survive arid environments in Africa. Euphorbs have undergone convergent evolution to the cacti of the Western Hemisphere.

Figure 9.95 Saguaro cactus, *Carnegiea gigantea*, is the largest of all North American cacti. Arms begin to develop on the saguaro when the plant is about 75 years old. A saguaro cactus may live over 250 years and reach a height of more than 50 feet.

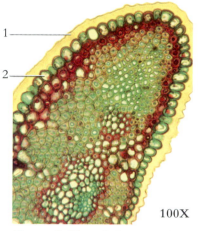

Figure 9.96 Prickly pear, *Opuntia* sp., cacti have several modifications to withstand drought. They have spine-like leaves to prevent water loss through transpiration; they have developed tissue that stores water after rain; and their stems are coated with a waxy substance to aid in water retention.

Figure 9.97 Fruit of the prickly pear, *Opuntia* sp.

Figure 9.98 Crown of thorns, *Euphorbia milii*, is native to Africa but is commonly cultivated in xeriscaped gardens.

Figure 9.99 Flowers from crown of thorns, *Euphorbia milii*.

Figure 9.100 Fishhook barrel cactus, *Ferocactus wislizeni*, expands in size during the rainy season in the American Southwest desert as it stores water.

Figure 9.101 Jumping cholla cactus, *Opuntia* sp., is well protected by dense spines. It is native to the American Southwest.

Figure 9.102 Cholla cacti, *Opuntia* sp., comprise many of the cacti species. Of the nearly 1,000 kinds of cacti, more than half are native to Mexico and the Southwestern United States.

Figure 9.103 Ocotillo, *Fouquiera* sp., in blossom during the rainy season in the Lower Sonoran desert. Small leaves extend from the stems during wet periods, but are quickly shed during arid periods.

Figure 9.104 Flowers of the Ocotillo, *Fouquiera* sp. Bats fertilize the blossoms of these flowers as they feed on the nectar. Ocotillo is often confused as being a cactus, however, it is a woody shrub.

Figure 9.105 Ball "mosses", *Tillandsia recurvata*, are not actually mosses, but flowering plants within the bromeliad family which are epiphytes (non-parasitic plants) that frequently grow on branches of various oak trees.

Figure 9.106 (a) Spanish "moss", *Tillandsia usneoides*, is actually a flowering plant related to the pineapple. (b) Detail of Spanish "moss", *Tillandsia usneoides*.

Flowers of Angiosperms

The angiosperm flower is composed of sepals, petals, stamens, and a carpel or carpels (*gynoecium*). The *sepals* are the outermost circle of protective leaf-like structures. They are green and are collectively called the *calyx*. The *petals* generally form a whorl inside of the calyx and are collectively called the *corolla*. Petals are often brightly colored and may secrete aromatic substances and nectar to attract pollinating insects. The *stamens* and the *carpels* are the reproductive parts of a flower. A stamen consists of the *filament* (stalk) and the *anther*, where *pollen* is produced. The centrally positioned *pistil* consists of a *stigma* at the tip that receives pollen and a *style* that leads to the *ovary*. The ovary is composed of one or more modified leaves known as carpels. A *carpel* is a megasporophyll upon which ovules are produced. The carpel encloses the ovules so that seeds are produced within a protective layer that matures to form a fruit. Most flowers contain both stamens and a pistil, although some species produce unisexual flowers.

On the basis of position of the ovaries, flowers are classified as *hypogynous* (with flower parts below the ovary), *epigynous* (with flower parts above the ovary), or *perigynous* (centrally-positioned ovary with floral parts on a cup-shaped receptacle). Regardless of the position of the ovary, most angiosperms rely on wind or animals for pollination. *Pollination* is the placement of pollen from the anther onto the stigma of the pistil by wind or animal vectors and is a prerequisite to fertilization. Wind is the primary pollinating agent for grasses and many trees. Because of this random dispersal, enormous quantities of pollen grains are released by the anthers of the flowers. Many angiosperms are pollinated by bees and other insects. The flowers of these plants are generally brightly colored and sweet smelling. Flowers pollinated by hummingbirds often have nectar deep in slender floral tubes where most other animals cannot reach.

When a *pollen grain* adheres to the stigma of the same species of plant, it swells and splits its outer coat. A tube cell grows and digests a tube down the style toward an ovule in the ovary. Directed by chemical attraction, the tip of the pollen tube enters the ovule and discharges two sperm nuclei into the embryo sac. One sperm fertilizes the *egg*, and the other combines with the two *polar nuclei* to form a triploid (3*n*) nucleus. The fusion of the egg with one sperm and the polar nuclei cell with the second sperm nucleus is called *double fertilization* and is unique to flowering plants. After fertilization, the other cells of the embryo sac degenerate and the ovule begins developing into a seed.

Flowers have contributed greatly to the success of angiosperms because they enhance the efficiency of plant reproduction by attracting and rewarding pollen-carrying animals. The beauty and fragrance of flowers have always appealed to humans. Even many perfumes have chemicals extracted from flowers as an important ingredient.

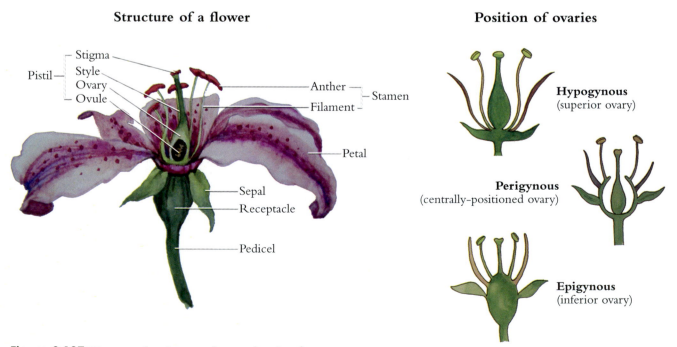

Figure 9.107 Diagram of angiosperm flowers showing the structure and relative position of the ovaries.

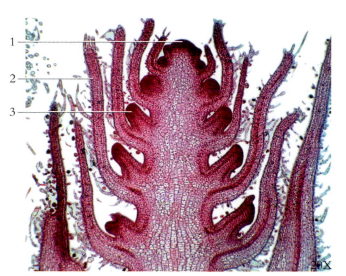

Figure 9.108 Floral bud of Coleus, *Coleus* sp.
1. Apical meristem   3. Floral bud
2. Bract

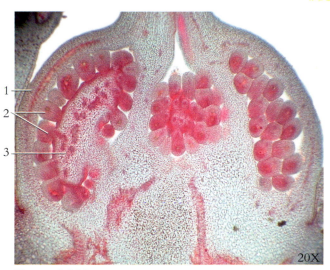

Figure 9.109 Ovary of tomato, *Lycopersicon* sp., with developing ovules
1. Ovary wall   3. Placenta
2. Ovules

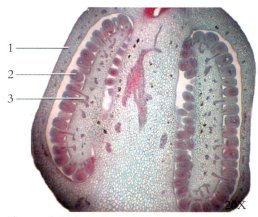

Figure 9.110 Nightshade, *Solanum* sp., floral bud showing ovary with developing ovules.
1. Ovary wall   3. Placenta
2. Ovules

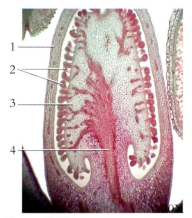

Figure 9.111 Floral bud of tobacco, *Nicotiana* sp., showing the ovary and ovules.
1. Ovary wall   3. Placenta
2. Ovules   4. Vascular tissue

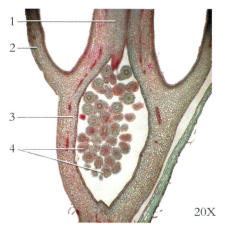

Figure 9.112 Floral bud of a currant, *Ribes* sp., showing an inferior ovary with developing ovules.
1. Style   3. Ovary
2. Petal   4. Ovules

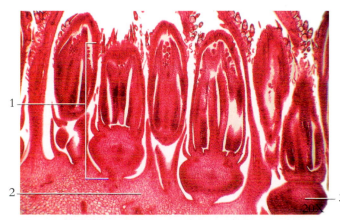

Figure 9.113 Floral bud of sunflower, *Helianthus* sp., with several immature flowers.
1. Individual flower   3. Ovary of individual flower
2. Receptacle

Figure 9.114 Floral structure of a tulip, *Tulipa* sp.

1. Petal   4. Filament
2. Anther   5. Ovary
3. Stigma

Figure 9.115 Structure of a dissected cherry, *Prunus* sp., showing a perigynous flower.

1. Petal   4. Anther   7. Floral tube
2. Filaments   5. Stigma
3. Sepal   6. Style

Figure 9.116 Structure of a dissected pear, *Pyrus* sp., showing an epigynous flower.

1. Petal   3. Filament   5. Sepal
2. Anther   4. Style   6. Ovary

Figure 9.117 Dissected quince, *Chaenomeles japonica*, showing an epigynous flower.

1. Petal   3. Stigma   5. Style
2. Anther   4. Filament   6. Ovules

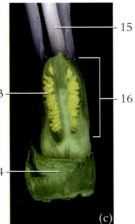

Figure 9.118 (a) The floral structure of *Gladiolus* sp. (b) The anthers and stigma and (c) the ovary.

1. Anther   10. Stigma
2. Filament   11. Style
3. Ovules   12. Filament
4. Receptacle   13. Ovules
5. Stigma   (immature seeds)
6. Style   14. Receptacle
7. Ovary   15. Style
8. Anther   16. Ovary
9. Pollen

Crown of Thorns

Foxglove

Morning Glory

Bird of Paradise

Wild Rose

Hibiscus

Chinese evergreen

Columbine

Orchid

Sunflower

Daisy

Dandelion

Figure 9.119 Flowers of representative angiosperms.

Figure 9.120 Scanning electron micrograph of the stigma of an angiosperm pistil. The stigma is the location where pollen grains adhere and germinate to produce a pollen tube.

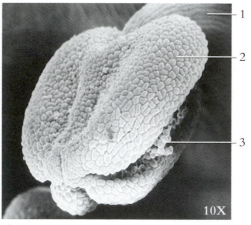

Figure 9.121 Scanning electron micrograph of the anther of candy tuft, *Lobularia* sp. The anther has ruptured, resulting in the release of pollen grains.
1. Filament　　　　3. Pollen grains
2. Anther

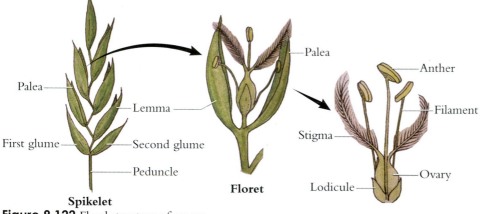

Palea

Lemma

First glume

Second glume

Peduncle

Spikelet

Palea

Floret

Anther

Filament

Stigma

Lodicule

Ovary

Figure 9.122 Floral structure of grasses.

Figure 9.123 Floral parts of a grass, *Elymus flavescens*, showing spikelets with six florets.

Figure 9.124 Grass, wheat, *Triticum* sp., is economically important.

Figure 9.125 Bamboo is an important grass of commerce. It is also important in many natural ecosystems.

Figure 9.126 Corn, *Zea mays,* is a New World grass important as food for humans and livestock.

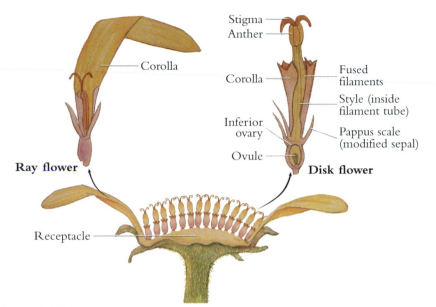

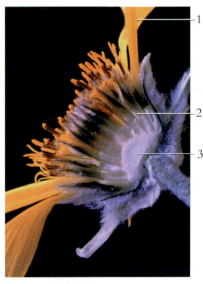

Figure 9.127 Flowers of the family Asteraceae are usually produced in tight heads resembling single large flowers. One of these inflorescences can contain hundreds of individual flowers. Examples of this family include dandelions, sunflowers, asters, and marigolds.

Figure 9.128 Dissected inflorescence of a member of the Asteraceae, *Balsamorhiza sagittata*.
1. Ray flower   3. Receptacle
2. Disk flower

Figure 9.129 Strawberry, *Fragaria* sp., showing (a) the flower, (b) immature aggregate fruits, and (c) a ripening fruit.

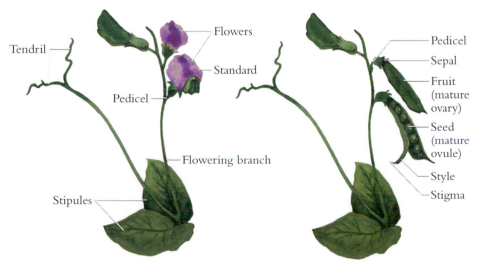

Figure 9.130 Flower and fruit of the pea, *Pisum* sp.

Figure 9.131 Fruits on the receptacle of the giant sunflower.

Figure 9.132 Lady slipper orchid, *Paphiopedilum* sp. The flower of the lady slipper orchid fills with rain and drops to the ground, allowing ants to enter and fertilize the flower.

Figure 9.133 Pollination of a flower by a hummingbird.

Figure 9.134 Flowers of many angiosperms are adapted for insect pollination, (a) through (d).

Angiosperm Reproductive Cycle

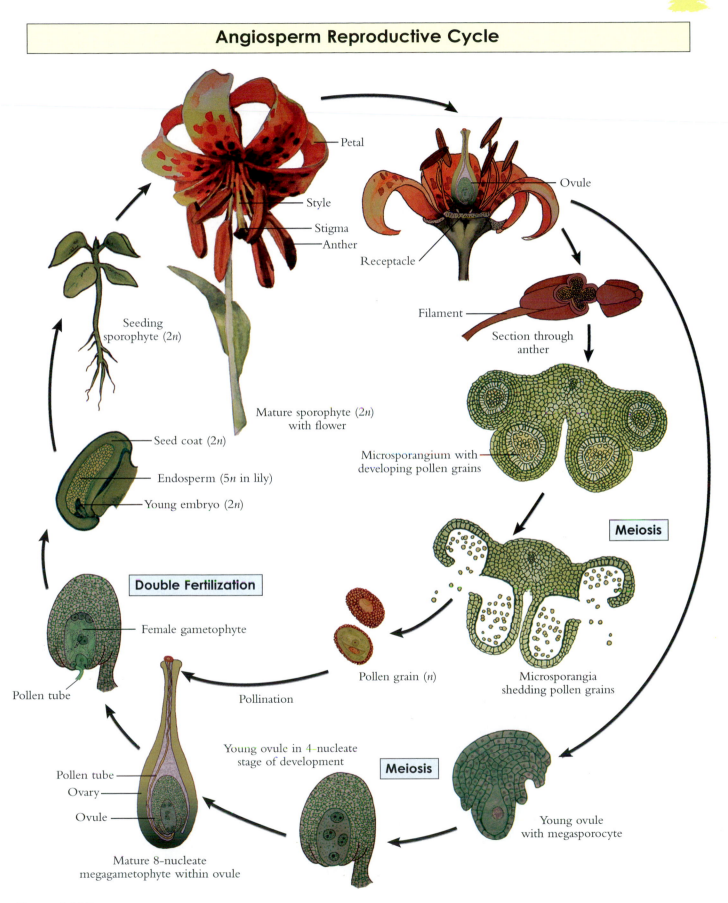

Figure 9.135 Life cycle of an angiosperm.

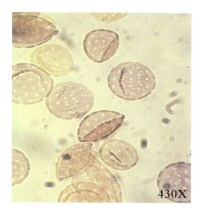

Figure 9.136 Pollen grains of the dicot pigweed, *Amaranthus* sp.

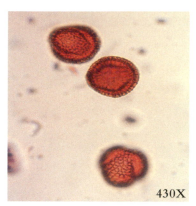

Figure 9.137 Pollen grains of a lilac, *Syringa* sp.

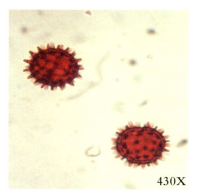

Figure 9.138 Pollen grains of the dicot arrowroot, *Balsamorhiza* sp.

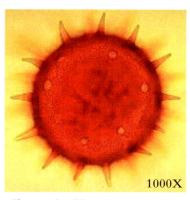

Figure 9.139 Pollen grain of hibiscus, *Hibiscus* sp.

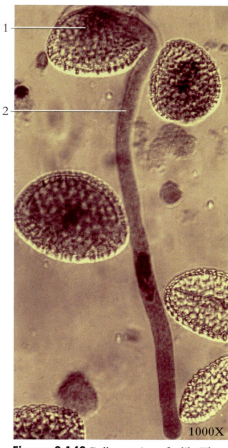

Figure 9.140 Pollen grains of a lily. The pollen grain at the top of the photo has germinated to produce a pollen tube.
1. Pollen grain   2. Pollen tube

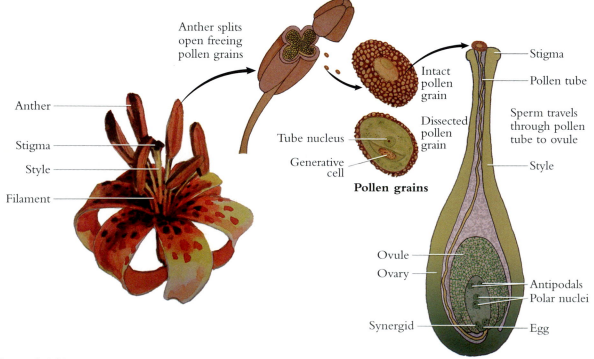

Figure 9.141 Diagram showing the process of pollination.

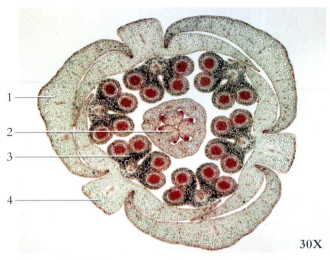

Figure 9.142 Transverse section of a flower bud from a lily, *Lilium* sp.

1. Sepal   3. Anther
2. Ovary   4. Petal

Figure 9.143 Transverse section of an anther from a lily, *Lilium* sp.

1. Sporogenous tissue   2. Filament

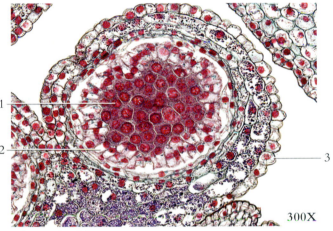

Figure 9.144 Transverse section of an anther from a lily, *Lilium* sp.

1. Young microsporocytes   2. Tapetum   3. Anther wall

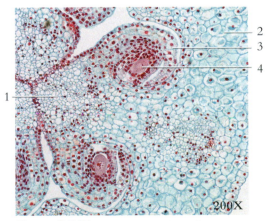

Figure 9.145 Transverse section of an anther from a lily, *Lilium* sp., magnified view.

1. Tapetum   2. Tetrad of microspores

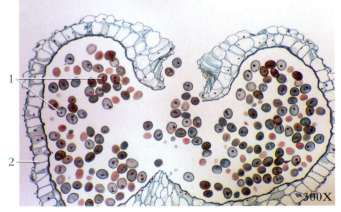

Figure 9.146 Transverse section of an anther from a lily, *Lilium* sp., showing mature pollen.

1. Pollen grains with two cells   2. Anther wall

Figure 9.147 Transverse section of a lily, *Lilium* sp., ovary showing ovules.

1. Placenta   3. Ovule
2. Ovary wall   4. Megasporocyte (*2n*)

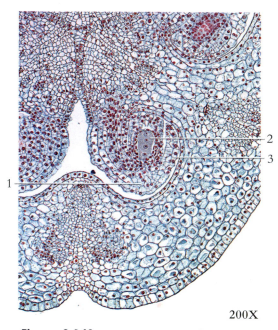

200X

Figure 9.148 Transverse section of a lily, *Lilium* sp., ovary showing megaspore.
1. Ovule
2. Linear tetrad of megaspore nuclei
3. Integument

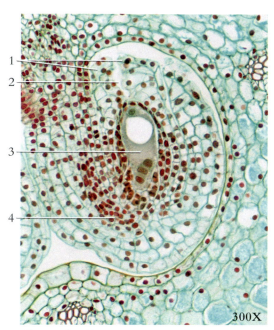

300X

Figure 9.149 Transverse section of a lily, *Lilium* sp., ovary showing ovule with developing embryo sac.

1. Integuments   3. Embryo sac
2. Micropyle   4. Ovule

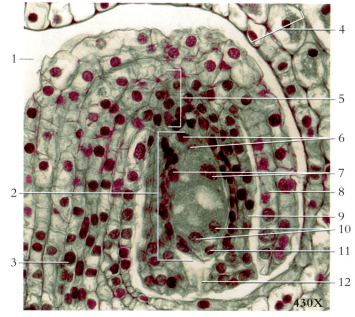

430X

Figure 9.150 Transverse section of an eight-nucleate embryo sac of an ovule from a lily, *Lilium* sp.

1. Locule
2. Megagametophyte
3. Funiculus
4. Wall of ovary
5. Chalaza
6. Antipodal cells ($3n$)
7. Polar nuclei ($1n$ and $3n$)
8. Outer integument ($2n$)
9. Inner integument ($2n$)
10. Synergid cells (n)
11. Egg (n)
12. Micropyle (pollen tube entrance)

200X

Figure 1.151 Photomicrograph of a mature grain, or kernel, of wheat, *Triticum aestivum*.

1. Pericarp
2. Starchy endosperm
3. Scutellum
4. Coleoptile
5. Shoot apex
6. Radicle
7. Coleorhiza
8. Embryo

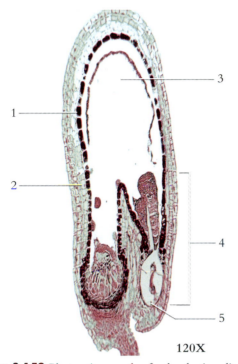

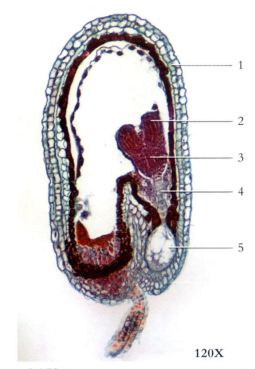

Figure 9.152 Photomicrograph of a developing dicot embryo from a shepherd's purse, *Capsella bursa-pastoris*.
1. Endothelium
2. Seed coat
3. Endosperm
4. Developing embryo
5. Basal cell

Figure 9.153 Photomicrograph of a developing dicot embryo from a shepherd's purse, *Capsella bursa-pastoris*, showing young embryo.
1. Seed coat
2. Cotyledon
3. Hypocotyl
4. Suspensor
5. Basal cell

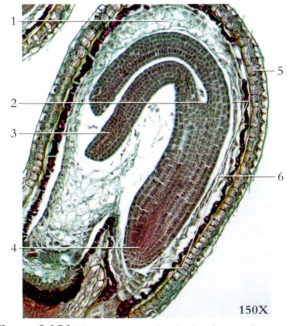

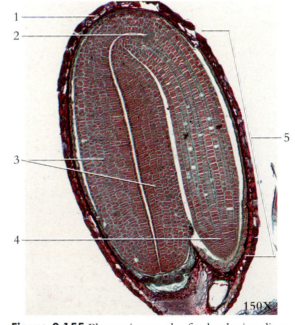

Figure 9.154 Photomicrograph of a developing dicot embryo from a shepherd's purse, *Capsella bursa-pastoris*, showing a nearly mature embryo.
1. Cellular endosperm
2. Epicotyl
3. Cotyledon
4. Radicle
5. Seed coat
6. Hypocotyl

Figure 9.155 Photomicrograph of a developing dicot embryo from a shepherd's purse, *Capsella bursa-pastoris*, showing a mature embryo..
1. Seed coat
2. Epicotyl
3. Cotyledons
4. Radicle
5. Hypocotyl

Seeds, Fruits and Seed Germination of Angiosperms

Seeds are the reproductive structures of gymnosperms and angiosperms. The seeds of gymnosperms (see Chapter 8) develop on the exposed surface of the scales of cones, whereas the seeds of angiosperms usually develop within a fruit produced from the ovary of a flower.

A typical seed of an angiosperm consists of a protective *seed coat*, a sporophyte *embryo*, and a layer of nutritive tissue called the *endosperm*. The endosperm, consisting of cells rich in proteins, fats, oils, and starch, is absorbed by the seed embryo during development, or germination. The embryo consists of *cotyledons, epicotyl, hypocotyl,* and *radicle*. During germination, the cotyledons become the embryonic leaves, the epicotyl becomes the shoot from which derives the first plant foliage, the hypocotyl is the point of attachment of the epicotyl, and the radicle becomes the primary root. When fully developed and prior to germination, some seeds dry out and become dormant with only 5 to 20% water content. A *fruit* develops around the angiosperm seed as it matures. Seeds and fruit are a source of food for many kinds of animals, including humans.

Although each species of angiosperm has evolved specific mechanisms for seed dispersal, three basic methods are used (see Figure 9.176).

1. *Animal-dispersed seeds* generally are produced in fleshy fruits (berries, grapes, cherries, apples) eaten by vertebrates. Seeds are dispersed unharmed as they are passed through the digestive tract. The enticing flavor and color of fruits are examples of coevolution of animals and flowering plants. Many other plants have fruits or seeds that have hooks, spines, or sticky coverings and are dispersed by adhering to fur or feathers.

2. *Water-dispersed seeds* include those from plants that grow near or in water and have seeds or fruit adapted for floating. In these species, either the seed or the fruit is buoyant. Nearly all Pacific islands have coconut trees that were seeded by buoyant coconuts. Rainfall is important in seed dispersal of some species.

3. *Wind-dispersed seeds* include those that are lightweight and buoyant in the air. The fruits of dandelions, for example, have dry, plume-like structures attached that carry the wind-borne fruits great distances. Each dandelion fruit contains one seed. Other plants, such as maples, develop fruits that dry into winglike structures. Tumbleweeds scatter their seeds as the detached plant blows along the ground. Other plants, such as the poppy, disperse their seeds aloft into the wind.

Fruits are classified on the basis of development and mature structure into three principal groups (see Figures 9.160-9.163): simple, aggregate, and multiple. *Simple fruits* develop from single ovaries and may be fleshy, such as cherries, or dry, such as legumes (beans and peas). *Aggregate fruits* develop from single flowers that have several separate carpels, such as strawberries and blackberries. *Multiple fruits* develop from groups of separate flowers clustered tightly, such as pineapples.

Seed germination occurs when appropriate environmental conditions are present. Some seeds must be exposed to extended cold; others must undergo a drying period followed by adequate moisture. Many seeds with hardened coats must be physically or chemically scarified before they can germinate. *Imbibition*, or the absorption of water, is the first step in the germination of most seeds. This *hydration* causes a seed to expand and rupture its coat. Once the germination process is initiated, the *radicle*, or root of the embryo, emerges from the seed and grows downward into the soil. In monocots, the shoot of the seedling grows upward through the tube of the *coleoptile*. In dicots, the hypocotyl of the seed grows upward, pulling the shoot and cotyledons from the soil. As leaves emerge from the seedling, the cotyledons die and wither, or may persist for weeks or months.

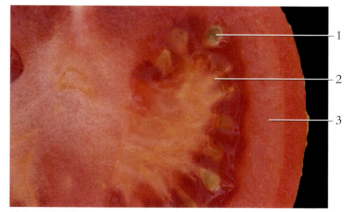

Figure 9.156 Mature seeds in the fruit of the tomato, *Lycopersicon esculentum.*
1. Seed
2. Placenta
3. Fruit wall

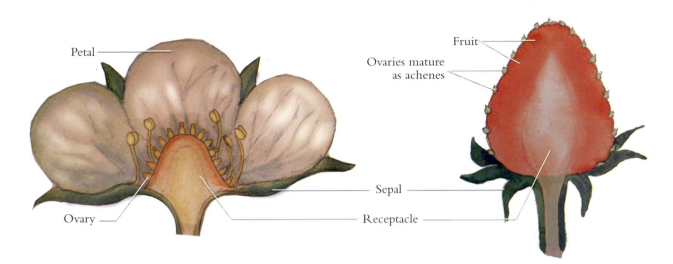

Figure 9.157 Flower and fruit of the strawberry, *Fragaria* sp. The strawberry is an aggregate fruit.

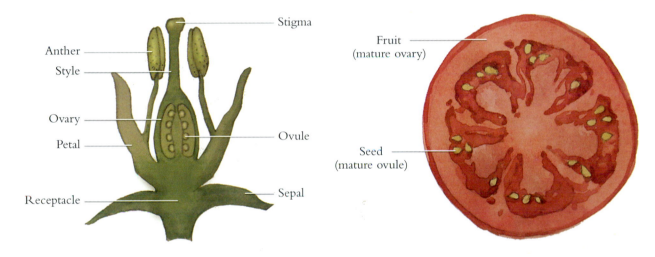

Figure 9.158 Illustration of a flower and fruit of a tomato, *Lycopersicon esculentum*. A tomato fruit is a berry.

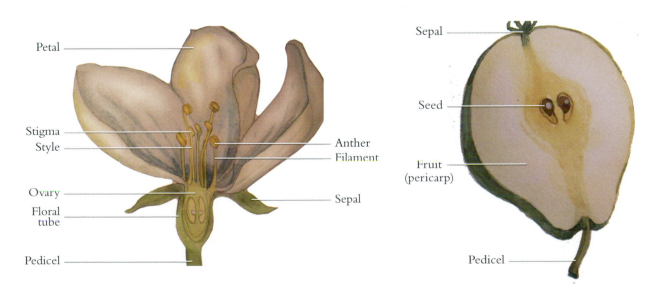

Figure 9.159 Flower and fruit of the pear *Pyrus* sp. The pear fruit develops from the floral tube (fused perianth) as well as the ovary.

(a) Drupe

(b) Berry

(c) Pome

(d) Drupe

(e) Legume

Figure 9.160 Examples of simple fruits: (a) peach, (b) grapes, (c) apple, (d) plum, and (e) pea.

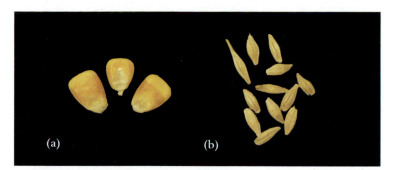

Figure 9.161 Examples of simple fruits: (a) corn and (b) oats.

Figure 9.162 Example of accessory fruit (a) strawberry and aggregate fruit (b) blackberry.

Figure 9.163 Examples of multiple fruits: (a) pineapple and (b) fig.

(a) (b) (c)

Figure 9.164 Flower (a) and the fruits (b and c) of the dandelion, *Taraxacum* sp. The dandelion has a composite flower. The wind-borne fruit (containing one seed) of a dandelion, and many other members of the family Asteraceae, develop a plumelike pappus, which enables the light fruit to float in the air.

1. Pappus   2. Ovary wall, with one seed inside

Figure 9.165 Dissected legume, garden bean, *Phaseolus* sp.

1. Pedicel   3. Fruit
2. Seeds   4. Style

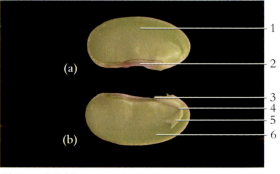

(a)

(b)

Figure 9.166 Lima bean. (a) The entire bean seed and (b) a longitudinally sectioned seed

1. Integument (seed coat)   4. Hypocotyl
2. Hilum   5. Epicotyl (plumule)
3. Radicle   6. Cotyledon

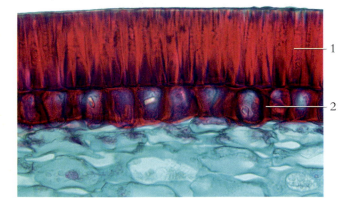

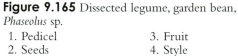

Figure 9.167 Photomicrograph of the seed coat of the garden bean, *Phaseolus* sp., showing the sclerified epidermis

1. Macrosclereids   2. Subepidermal sclereids

Figure 9.168 Cob of corn from *Zea mays*. Corn was domesticated approximately 7,000 years ago from a Mexican grass, family Poaceae.

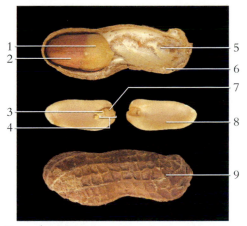

Figure 9.169 Fruit and seed of a peanut plant.

1. Cotyledon
2. Integument (seed coat)
3. Plumule
4. Embryo axis
5. Interior of fruit
6. Mesocarp
7. Radicle
8. Cotyledon
9. Fruit wall (pericarp)

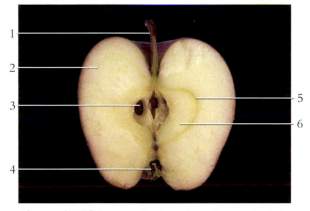

Figure 9.170 Longitudinal section of an apple fruit.

1. Pedicel
2. Mature floral tube
3. Seed (mature ovule)
4. Remnants of floral parts
5. Ovary wall
6. Mature ovary (2 & 6 comprise the fruit)

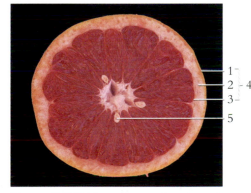

Figure 9.171 Transverse section through a grapefruit fruit.

1. Exocarp
2. Mesocarp
3. Endocarp
4. Pericarp
5. Seed

Figure 9.172 Longitudinal section of a pineapple fruit.

1. Shoot apex   2. Central axis   3. Floral parts

Figure 9.173 Longitudinal section of a tomato fruit (berry).

1. Pedicel
2. Pericarp
3. Locule
4. Placenta
5. Seed
6. Sepals
7. Mature ovary (fruit)

Figure 9.174 Examples of seed dispersal.

(a) Maple—The winged fruits of a maple fall with a spinning motion that may carry it hundreds of yards from the parent tree.
(b) White pine—The second-year cones of a white pine open to expose the winged seeds to the wind.
(c) Willow—The airborne seeds of a willow may be dispersed over long distances.
(d) Witch hazel—Mature seeds of the witch hazel tree are dispersed up to 10 feet by forceful discharge.
(e) Mangrove—The fruits of this tropical tree begin to germinate while still on the branch, forming pointed roots. When the seeds drop from the tree, they may float to a muddy area where the roots take hold.
(f) Coconut—The buoyant, fibrous husk of a coconut permits dispersal from one island or land mass to another by ocean currents.

(g) Pecan—The fruit husk of a pecan provides buoyancy and protection as it is dispersed by water.
(h) Black walnut—The encapsulated seed of the black walnut is dispersed through burial by a squirrel or floating in a stream.
(i) Apple—The seeds of an apple tree may be dispersed by animals that ingest the fruit and pass the undigested seeds hours later in their feces.
(j) Cherry—Moderate-sized birds, such as robins, may carry a ripe cherry to an eating site where the juicy pulp is eaten and the hard seed is discarded.
(k) Beech—Seeds from a beech tree are dispersed by mammals as the spiny husks adhere to their hair. In addition, many mammals ingest these seeds and disperse them in their feces.
(l) Oak—An oak seed may be dispersed through burial of the acorn fruit by a squirrel or jay.

Figure 9.175 (a) Mature milkweed, *Asclepias* sp.; (b) milkweed pods; and (c) seeds ready for airborne dispersal.

Forcible discharge dispersal

Touch-me-not

Water dispersal

Coconut

Animal dispersal

Burdock   Cocklebur

Blackberries

Wind dispersal

Dandelion   Poppy

Maple

Figure 9.176 Several fruits and seeds to illustrate seed dispersal.

Figure 9.177 The duckweed, *Lemna* sp., is a small free-float-ing fresh-water plant found throughout the United States. The flowers are small and unisexual. Within the family Lemnaceae, *Lemna* is one of the smallest flowering plants in the world (scale in mm.).

Figure 9.178 *Euphorbia canariensis*, a member of the spurge family, Euphorbiaceae. Living in African and Australian desert environments, species of *Euphorbia* provide examples of convergent evolution with specimens of the New World cacti of the family Cactaceae.

Representative Specimens of Angiosperms

Figure 9.179 Herbarium specimen of a sage, *Salvia dorrii*, family Lamiaceae. *Salvia* lives in arid environments where it produces terpenes that inhibit the growth of other plants. This specialization tends to insure adequate moisture for an established plant.

Figure 9.180 Herbarium specimen of loco weed, *Astragalus oophorus,* family Fabaceae. Loco weed is toxic to livestock on the semiarid open ranges in the Western United States.

Figure 9.181 Herbarium specimen of papyrus, *Cyperus papyrus*. *Cyperus* is within the family Cyperaceae. Papyrus, a tropical reed that grows in the water ways of Northern Africa, was used by the Egyptians to make paper.

Figure 9.182 Herbarium specimen of Indian rice grass, *Stipa hymenoides*. Indian rice grass is a member of the family Poaceae and was used by Native Americans to make a mush-like food, rich in protein. *Stipa hymenoides* is the state grass of both Utah and Nevada.

Figure 9.183. Herbarium specimen of the sedge, *Carex scirpoidea,* family Cyperaceae. This species of sedge is a high altitude plant that occurs above the timberline in North America. The higher the altitude, the smaller the plant specimens become.

Figure 9.184 Herbarium specimen of a wild rose, *Rosa woodsii*. With only five petals, *Rosa woodsii* is an ancestral form of cultivated roses within the rose family, Rosaceae.

Figure 9.185 Herbarium specimen of holly atriplex, *Atriplex hymenelytra*. Holly atriplex is a salt-tolerant plant within the family Chenopodiaceae. This species of atriplex has small, felt-like leaves. Its range is arid to semiarid regions in the Western United States.

Figure 9.186 Herbarium specimen of a lady slipper, *Cypripedium calceolus*, family Orchidaceae. The lady slipper orchid is found in wet climates in the eastern United States and in mountainous regions of other parts of the United States.

Figure 9.187 Herbarium specimen of breadroot, *Cymopterus purpurascens*. Breadroot is a member of the family Apiaceae. Developing foliage in the spring, this plant was used as a food by many Native Americans.

Figure 9.188 Herbarium specimen of *Verbascum thapsus*. An Old World plant within the family Scrophulariaceae, *Verbascum thapsus* was used in making fire torches. Portions of the plant were also used for medicinal purposes.

Figure 9.189 Herbarium specimen of the small barrel cactus, *Neolloydia johnstonii*, family Cactaceae. This cactus is endemic to the western United States and is noted for its brilliant purplish-pink flower.

Figure 9.190 Herbarium specimen of *Cercocarpus montanus*, family Rosaceae. The wood of this plant is extremely dense and was utilized by many Native Americans for making bows and arrows.

Figure 9.191 Herbarium specimen of a sunflower, *Helianthus annuus*. Sunflowers produce flowers in a composite inflorescence and are members of the family Asteraceae.

Figure 9.192 Herbarium specimen of rice, *Oryza sativa*. A species within the grass family, Poaceae, *Oryza sativa* is the major food crop grown in Asia. Known to have been cultivated for over 7,000 years, rice requires warm temperatures and abundant moisture. Rice is one of the twelve most important human food plants.

Figure 9.193 Herbarium specimen of wheat, *Triticum aestivum*. Like the other cereal crops, wheat belongs to the grass family, Poaceae. Wheat was first cultivated in the Middle East over 9,000 years ago. It is currently grown in temperate climates throughout the world and is one of the twelve most important human food plants.

Figure 9.194 Herbarium specimen of barley, *Hordeum vulgare*. Like wheat, rye is a member of the grass family, Poaceae. Rye has been cultivated as a grain crop since the time of ancient Rome. It is currently grown in cool climates of northern Europe, Asia, North America, and South America.

Figure 9.195 Herbarium specimen of maize (corn), *Zea mays*. Domesticated nearly 7,000 years ago from a Mexican grass in the family Poaceae, maize is currently grown throughout the world but more extensively in North America. During pre-Columbian times, it was cultivated by Native Americans in societies throughout North and South America. Although more than half of the cultivated maize in the United States is used for animal feed, it is the basis of many important food items ranging from corn bread to cereals to tortillas. Maize is one of the twelve most important human food plants.

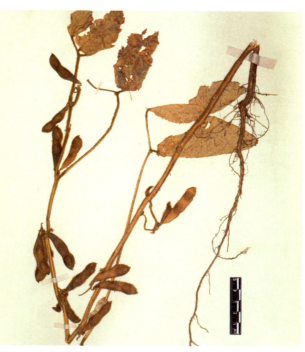

Figure 9.196 Herbarium specimen of the soybean, *Glycine max*. A legume within the family Fabaceae, the soybean was initially cultivated in China nearly 5,000 years ago. Currently more than half of the world's soybean production is in the United States. Soybeans and other beans are high in nutrients and easily grown in many parts of the world. Because of this, the soybean is considered one of the 12 most important human foods.

Figure 9.197 Herbarium specimen of cassava, or manioc, *Manihot esculenta*. Within the spurge family Euphorbiaceae, cassava is an important starch root crop in South America, West Indies, Africa, and Indonesia. Cassava is a shrub that has large, starch-filled roots. Following the preparation of the root, the residue may be utilized in many ways, including baking as thin cake bread or drying as a meal called farinha. It is also the source of tapioca, used in puddings. Cassava is considered one of the twelve most important human food plants.

Figure 9.198 Herbarium specimen of a garden bean plant, *Phaseolus vulgaris*. The garden bean is easily grown and the pod and seeds of the fruit provide a nutritious vegetable.

Figure 9.199 Herbarium specimen of the white potato, *Solanum tuberosum*. The potato is a member of the family Solanaceae. The potato tuber is a modified stem that is rich in nutrients and is a stable crop for millions of people. Cultivated by Native South Americans, potatoes were introduced to Europe about 1570. Potatoes are currently grown worldwide, especially in temperate regions and at higher tropical elevations. The potato is one of the twelve most important human food plants.

Figure 9.200 Herbarium specimen of the coconut, *Cocos nucifera*. *Cocos nucifera* is a species within the palm family Arecaceae (or Palmae). Distributed along tropical shorelines, the coconut "meat" is a nutritious source of food for millions of people. Palm leaves are used in making shelters and the fibrous portion of the coconut fruit is used for making mats and rope. The coconut is one of the twelve most important human food plants.

Figure 9.201 Herbarium specimen of the banana, *Musa balbisiana*. There are more than 300 varieties of bananas, all within the family Musaceae. Thought to have been domesticated thousands of years ago, the first became known to Europeans following the incursion of Alexander the Great into India (327 b.c.). Bananas are high in nutritional value. The banana is one of the twelve most important human food plants.

Figure 9.202 Herbarium specimen of a squash plant, *Cucurbita foetidissima*, family Cucurbitaceae. Many varieties of squash have been cultivated and are important food crops.

Figure 9.203 Herbarium specimen of tobacco, *Nicotiana tabacum*. The tobacco plant is in the nightshade family Solanaceae. Native to the American tropics, tobacco was first encountered by Columbus and his men in the West Indies. It is currently cultivated in countries throughout the world and as many as half the world's population chew, sniff, or smoke the products from the tobacco plant for the effects of the nicotine.

Figure 9.204 Herbarium specimen of sugarcane, *Saccharum ravennae*. There are several species of cultivated sugarcane grown in moist tropical regions throughout the world. Initially used in India nearly 5,000 years ago, sugarcane is a sturdy perennial grass with broad leaves. The sucrose content within the stems of certain varieties may reach 20% of the crop biomass. Sugarcane is one of the twelve most important human food plants.

Figure 9.205 Herbarium specimen of the hot pepper, *Capsicum frutescens*. Native to North America, the hot pepper was introduced to Europe by Columbus. Because of its value in spicing foods, it is an important cultivated plant.

Figure 9.206 Herbarium specimen of the hemp plant, *Cannabis sativa*. The hemp plant is the source of marijuana and hashish. It is a dioecious annual that was initially cultivated in China as early as 3,000 b.c. The seeds of hemp are used for industrial oil and the plant body is a source of valuable fiber. In spite of its cultivation being illegal in many countries, products from the hemp plant are used by an estimated 200 million people throughout the world, making it an important cultivated crop.

Figure 9.207 Herbarium specimen of the opium poppy, *Papaver somniferum*. The latex of opium contains many alkaloids, of which morphine and codeine are the most important because of their medicinal benefits in making pain-killers. As a cultivated plant for the narcotics it produces, the opium poppy was first grown in Asia Minor as early as 2,500 b.c. Opium and its derivatives are commonly used narcotics, with an estimated 900 million users, mostly in Asia.

Figure 9.208 Herbarium specimen of the foxglove, *Digitalis purpurea*, within the family Plantaginaceae. Foxglove is a native European wild flower that is an important medicinal plant to alleviate cardiac insufficiency and related problems.

Angiosperms of Commercial Value as Wood Products

Red alder–*Alnus rubra*

Linden (basswood)–*Tilia americana*

Ash

Red alder–*Alnus rubra*.
Principal hardwood in the forests of the Pacific Northwest. Size 40 to 50 feet in height. Commercially important for construction of cabinets and furniture; frames and doors, veneer for plywood, pulpwood, and charcoal.

White ash–*Fraxinus americana*.
Distribution eastern North America. Size 80 to 100 feet in height. Commercially important for construction of tool handles (shovels, rakes, hoes), sporting equipment (baseball bats, hockey sticks, paddles, and oars), wooden toys, and cabinets.

Basswood

Linden (basswood)–*Tilia americana*.
Distribution eastern North America. Size 80 to 90 feet in height. Commercially important for construction of building materials: plywood, furniture, cabinets, and doors; piano keys, boxes, and caskets.

Beech

American beech–*Fagus americana*.
Distribution eastern North America. Size 70 to 80 feet in height. Commercially important for construction of building materials: flooring and veneer; fuel wood; chemicals such as acetic acid and methanol.

Birch

Yellow birch–*Betula alleghaniensis*.
Distribution eastern North America. Size 80 to 100 feet in height. Commercially important for construction of furniture, veneer, musical instruments; boxes and matches.

Cherry

Black cherry–*Prunus serotina*.
Distribution throughout wooded areas of North America. Size 70 to 80 feet in height. Commercially important for construction of cabinets, veneer for plywood, flooring, furniture, paneling, and interior trim.

Cottonwood

Cottonwood–*Populus deltoides*.
Distribution throughout most of northern United States. Size 75 to 90 feet in height. Commercially important for construction of pulpwood for paper, draws, pails, boxes, and crates.

Elm

American elm–*Ulmus americana*
Distribution eastern United States and southeastern Canada. Size 75 to 100 feet in height. Commercially important for construction of furniture, crates, boxes, and round cheese containers.

Hickory

Bitternut hickory–*Carya cordiformis*.
Distribution eastern North America. Size 80 to 1000 feet in height. Commercially important for construction of furniture and tool handles; fuel wood and smoking meat.

Locust

Black locust–*Robinia pseudo-acacia*
Distribution southeastern United States. Size 70 to 80 feet in height. Commercially important for construction of wooden bridges and planks, mine timber, railroad ties, and fence posts.

Maple

Sugar maple–*Acer saccharum*.
Distribution eastern North America. Size 80 to 100 feet in height. Commercially important for construction of furniture, veneer for plywood, flooring, woodenware, and musical instruments. Sap is processed into maple extract and maple syrup.

Oak

Red oak–*Quercus rubra*.
Distribution eastern United States. Size 70 to 80 feet in height. Commercially important for construction of flooring, doors, boat-building, and caskets; fuel wood.

White oak–*Quercus alba*.
Distribution eastern North America. Size 70 to 120 feet in height. Commercially important for construction of flooring, doors, furniture, boat-building, and caskets; fuel wood; barrels for aging wine and whiskey.

Persimmon

Persimmon–*Diospyros virginiana*.
Distribution eastern United States. Size 80 to 100 feet in height. Commercially important for construction of furniture, boats, barrels, and golf-club heads.

Sycamore

American sycamore–*Platanus occidentalis*.
Distribution eastern United States. Size 70 to 100 feet in height. Commercially important for construction of interior trim, paneling, flooring and draws; butcher's blocks.

Walnut

Black walnut–*Juglans nigra*.
Distribution central and eastern United States. Size 80 to 100 feet in height. Commercially important for construction of cabinets, veneer for plywood, chairs, tables, and furniture; gunstocks and coffins.

Poplar

Yellow poplar (tulip tree)–*Liriodendron tulipifera*.
Distribution eastern North America. Size 140 to 150 feet in height. Commercially important for construction of furniture, doors, boxes, and veneer for plywood.

White oak–*Quercus alba*

American sycamore–*Platanus occidentalis*

Photos of Live Specimens of Angiosperms

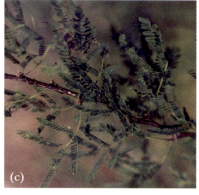

Figure 9.209 Acacia (*Acacia*). Native to the southwestern United States, this species of acacia commonly occurs along the stream banks of gullies and washes. It ranges in height from 25-feet to 35-feet (a). The fruit of acacia are contained in pods that split open to release the mature seeds, which are eaten by many species of birds and small mammals. The bark (b) is rough and leaves (c) small to retain moisture.

Figure 9.210 Crapemyrtle (*Lagerstroemia indica*). Native to Japan and eastern Asia, these flowering trees are cultivated for landscaping in southwestern United States. They frequently grow to a height of 40 feet (a) and have distinctive bark (b) and glossy, green leaves (c).

Figure 9.211 Burr oak (*Quercus macrocarpa*). Native to eastern North America, this rounded crown tree (a) attains heights of 60 to 80 feet and has rough bark (b). The obovate leaves (c) are deeply furrowed into round-ended lobes and are glossy green in color. The acorn fruit is 2 inches long and is an important food source for rodents and many species of birds.

Figure 9.212 Canby oak (*Quercus canbyi*). Native to the mountains of northern Mexico, this semi-deciduous tree (a) grows to about 35 feet. The tree has rough bark (b), glossy, serrated, dark green leaves (c), and is drought tolerant. Frequently used for firewood, this tree is adapted to cold temperatures well below freezing.

(a)

(b)

(c)

Figure 9.213 Hill country live oak (*Quercus fusiformis*) This evergreen oak (a) is native to Texas, particularly in hilly terrain. It grows to about 80 feet tall and develops a crown of 70-feet to 100-feet wide. The bark is grayish-brown (b) and furrowed. The leaves (c) of this drought-tolerant tree are dark green and shiny. The branches of this tree are frequently covered with ball moss, which is a harmless epiphyte (non-parasitic) bromeliad. The wood is used for furniture.

(a)

(b)

(c)

Figure 9.214 Coastal live oak (*Quercus virginiana*) Common along the coastal states of southeastern United States, this tree (a) grows to a height of about 50 feet. Its elliptical, leathery leaves (c) are evergreen. The branches are frequently covered with Spanish moss, an epiphytic bromeliad. The bark is furrowed (b). The wood of this tree is utilized in making furniture and oak cabinets.

Figure 9.215 Chinkapin oak (*Quercus muehlenbergii*). This moderately fast growing tree (a) is native to the eastern deciduous forest, Texas, and eastern New Mexico. Its heights range from 65 feet to 100 feet. The bark is grayish-brown (b) and furrowed. The dense wood is used for railroad ties, posts, construction timber, and fuel. Leaves are drought-tolerant (c). Acorns from this tree are only about 1/2 inch long.

Figure 9.216 Lacey oak (*Quercus laceyi*). Native to the mountains of Mexico and the hills of southern Texas, this drought-tolerant oak (a) grows to about 45 feet with a crown spread of 30 feet. The bark brown (b) and rough. The leaves (c) are grayish-green that change to yellow in the fall prior to dropping.

Figure 9.217 Gambel oak (*Quercus gambelii*). Native to western North America, this rounded crown tree (a) attains heights of 40 to 60 feet. The obovate leaves (c) are deeply furrowed into round-ended lobes and are glossy green in color. The bark is grayish-brown (b). The acorn fruit is 1.5 inches long and is an important food source for rodents and many species of birds.

Figure 9.218 Silver maple (*Acer saccharinum*). A large tree (a) that grows to a height of 60 feet to 90 feet. Older bark tends of flake (b), leaving brown spots. Leaves (c) are deeply five-lobed and silvery-white.

Figure 9.219 Sweet gum (*Liquidambar styraciflua*). This 80-foot to 120-foot tree is native to southeastern United States and along damp coastal plains (a). Leaves (c) are deeply palmate and 5-lobed. Its small yellow-green flowers are without petals, and its hanging fruit is enclosed in a ball covered with tiny horns (d). The wood of the sweet gum is used for furniture, cabinets, and veneer.

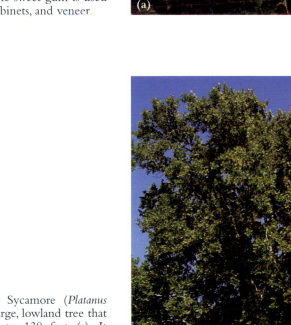

Figure 9.220 Sycamore (*Platanus occidentalis*). A large, lowland tree that grows 75 feet to 130 feet (a). It has distinctive mottled brown bark (c) that frequently flakes off in a jigsaw-puzzle-like pattern exposing the yellowish underbark. Its hanging fruit is enclosed in a ball covered with tiny horns (b). It is native to eastern United States. Its dense, coarse-gained wood is used for boxes, barrels, butchers' blocks, cabinetwork, and furniture. Native Americans frequently hollowed out the trunks of sycamores to use them as canoes.

Figure 9.221 Catalpa (*Catalpa bignonioides*). Catalpa grows to a height of 50 to 70 feet (a) and has large (6" to 12") leaves (c). Flowers are white with yellow and purple spots. The bark is grayish-brown (b). Fruits are long slender pods (d) containing many small seeds.

Figure 9.222 Russian olive (*Elaeagnus angustifolia*) (a). A moderately sized, bushy tree with reddish-gray bark (b). Leaves (c) are elliptical and silver brown in coloration. Flowers are silvery yellow and fragrant. Fruits are silvery, small, and elliptical. Russian olive trees are one of a few non-legumes that fix nitrogen in the soil in bacterial root nodules.

Figure 9.223 Linden (*Tilia americana*). Distribution eastern North America. Size 80 to 90 feet in height. The linden has a triangle shape (a). The bark is gray-brown (b). The leaves are one-sided, and heart-shaped (c). Commercially important for construction of building materials: plywood, furniture, cabinets, and doors; piano keys, boxes, and caskets.

Figure 9.224 American hornbeam (*Carpinus caroliniana*). Also known as blue beech or water beech, this tree (a) is native to eastern North America. As a mature tree of about 30 feet in height, it is somewhat circular in shape. The bark is smooth and gray (b). Its oval leaves (c) are double toothed and dark green. They change to an orange to red color in autumn. The dense and heavy wood of this tree is used for tool handles and wedges.

Figure 9.225 Bluewood condalia (*Condalia hookeri*). This drought tolerant spiny shrub or small tree (a) is native to the southwest United States and northwestern Mexico. The bark is smooth and gray (b). Leaves are small and waxy (c). The dark blue fruit is eaten by many desert birds and mammals and makes good jelly and wine.

Figure 9.226 Yaupon holly (*Ilex vomitoria*). Native to south eastern and central Texas to northern Mexico, this 30-foot tree (a) has dark glossy evergreen leaves and red berries (c) (on females) that persist through the winter. The bark is smooth and gray (b). Many species of birds feed on the berries. Native Americans used the berries to make a beverage called "black drink" that was used in a ritualistic purging ceremony.

A

Abiotic: that portion of the environment without living organisms; the non-living portion of the environment.

Abscisic acid: a plant hormone that inhibits growth and promotes dormancy; helps the plant conserve energy and withstand stressful conditions.

Abscission: the shedding of leaves, flowers, fruits, or other plant parts, usually following the formation of an abscission zone.

Absorption: movement of a substance into a cell or an organism, or through a surface within an organism.

Accessory bud: a bud developing on a stem or twig located above or on either side of the axillary bud.

Accessory fruit: a fruit composed primarily of tissue other than ovary tissue. Apples, plums, and pears are examples of accessory fruit.

Acclimation: the structural and functional changes of an organization in response to climatic changes.

Acid: a substance that releases hydrogen ions in a solution.

Acid rain: acidic atmospheric water formed from sulfur and/or nitrogen oxides forming acids when they react with water; partially due to the combustion of coal.

Actinomorphic: a flower type that can be divided into halves in more than one longitudinal plane (also called *radially symmetrical plane*).

Active transport: energy-requiring movement of a molecule across a membrane from a region of low concentration to a region of high concentration.

Adaptive radiation: the emergence of several species from a common ancestor.

Adhesion: an attachment between unlike substances or cells. No chemical bonds are formed between the two.

Adventitious root: a root developing from the stem of a plant; often functioning in support.

Aeciospore: a dikaryotic spore of rust fungi.

Aecium: a cuplike structure in a rust fungus where aeciospores are produced.

Aerobic: respiration requiring free oxygen (O_2) for growth and metabolism.

Aggregate fruit: a fruit produced from a single flower with several separate carpels.

Alga (*pl. algae*)**:** any of a diverse group of aquatic photosynthesizing organisms that are either unicellular or multicellular; algae comprise the phytoplankton and seaweeds of the Earth.

Alkaline: a substance having a pH greater than 7.0; *basic.*

Allele: an alternative form of gene occurring at a given chromosome site, or locus; several alleles may exist for a single gene.

Alternation of generations: two-phased life cycle characteristic of many plants in which sporophyte and gametophyte generations alternate.

Amoeba: a protozoan that moves by means of pseudopodia.

Anaerobic: metabolizing and growing in the absence of free oxygen (O_2).

Analogous: structures that have a common function or appearance in different species but lack a common developmental origin.

Anatomy: the structure of an organism.

Angiosperm: flowering plant, having double fertilization resulting in development of specialized seeds within fruits.

Annual plant: a plant that completes its entire life cycle in a single year or growing season.

Annual ring: yearly growth demarcation in woody plants formed by production of "spring wood" and "summer wood" in the secondary xylem.

Anther: the terminal pollen sacs of a stamen in an angiosperm flower where pollen grains with male gametes develop.

Antheridium (*pl. antheridia*)**:** a sperm-producing structure in an alga and some fungi.

Apical meristem: embryonic plant tissue in the tip of a root, bud, or shoot where continual cell divisions cause growth in length.

Archaea: one of the three domains of organisms that includes methanogens, halophiles, and thermophiles.

Archegonium: multicellular female reproductive organ in certain plants; a gametangium where eggs are produced.

Ascospore: a haploid spore produced within an ascus of a sac fungus (ascomycete).

Asexual reproduction: a reproduction process that does not require the union of gametes, such as budding or fission.

Asymmetry: a non-symmetrical morphology.

Autosome: a chromosome other than a sex chromosome.

Autotroph: an organism capable of synthesizing organic molecules (food) from inorganic molecules.

Auxin: a category of plant hormone that stimulates cell differentiation and plant growth, such as phototropic response through cell elongation, stimulation of secondary growth, and development of leaf traces and fruit.

Axillary bud: group of meristematic cells at the junction of a leaf and stem, which may develop a branch or flower(s); also called *lateral bud.*

B

Bacillus *(pl. bacilli)*: a rod-shaped bacterium.

Bacterium *(pl. bacteria)*: a prokaryote within the Bacteria domain, lacking the organelles of an eukaryotic cell.

Bark: outer tissue layers of a woody plant consisting of periderm, cortex, and outer phloem.

Basal: at or near the base or point of attachment.

Basidium *(pl. basidia)*: a reproductive cell of basidiomycetes, where nuclear fusion to form a diploid cell followed by meiosis occurs to produce basidiospores.

Berry: a simple fleshy fruit that develops from a superior ovary wall.

Biennial plant: a plant that lives through two growing seasons; generally, these plants often have vegetative growth during the first season, and flower and set seed during the second season.

Bilateral symmetry: the morphologic condition of having similar halves.

Binomial nomenclature: an assignment of two names to an organism, the first of which is the genus and the second the specific epithet; together constituting the *scientific name.*

Biomass: the dry weight of all organisms in a particular sample or area.

Biome: a major climax community characterized by a particular group of plants and animals.

Biosphere: the portion of the Earth's atmosphere and surface where living organisms exist.

Biotic: pertaining to the living part of the environment.

Bisexual flower: a flower that produces both male and female sex organs.

Blade: the broad expanded portion of a leaf.

Brackish: water that is intermediate in saltiness between fresh water and seawater.

Bryophyte: a plant within the phylum Bryophyta; a moss, liverwort, or hornwort; non-vascular plant that inhabits terrestrial environments but lacks many of the adaptations of most vascular plants.

Budding: type of asexual reproduction in which outgrowths from the parent plant pinch off to live independently or may remain attached to form colonies.

Bulb: a thickened underground stem often enclosed by enlarged, fleshy leaves containing stored food.

C

Callus: a mass of undifferentiated plant tissue often growing in a wound during the healing process.

Carpel: the megasporophyll of an angiosperm.

Carrying capacity: the maximum number of organisms of a species that can be maintained indefinitely in an ecosystem without causing damage.

Catalyst: a chemical, such as an enzyme, that accelerates the rate of a reaction of a chemical process but is not used up in the process.

Cell: the structural and functional unit of an organism; the smallest structure capable of performing all the functions necessary for life.

Cell wall: a rigid protective structure of a plant cell surrounding the cell (plasma) membrane; often composed of cellulose fibers embedded in a polysaccharide/protein matrix.

Cellular respiration: the reactions of glycolysis, Krebs cycle, and electron transport system that provided cellular energy and accompanying reactions to produce ATP.

Cellulose: a polysaccharide produced as fibers that form a major part of the rigid cell wall around a plant cell.

Chlorophyll: green pigment in photosynthesizing organisms that absorbs energy from the sun.

Chloroplast: a double membrane-enclosed organelle that contains chlorophyll and is the site of photosynthesis.

Chromosome: structure in the nucleus of a cell that contains the genes; comprised of a molecule of DNA and associated proteins.

Climax community: a mature biological community that is the relatively stable terminal stage reached in ecological succession.

Colony: an aggregation of organisms of the same species living together in close proximity.

Community: an ecological unit composed of all the populations of organisms living and interacting in a given area.

Competition: interaction between individuals of the same or different species striving to obtain a mutually necessary resource.

Complete flower: a flower that has the four types of floral components including sepals, petals, stamens, and carpels.

Compound leaf: a leaf with a blade deeply divided into distinct leaflets.

Conifer: a cone-bearing woody seed plant, such as pine, fir, and spruce.

Convergent evolution: the evolution of similar structures in different groups of organisms. They lack common ancestors but occur in similar environments.

Cork: the protective outer layer of bark of many trees, composed of dead cells that may be sloughed off.

Cortex: a primary tissue region of a plant root or stem bounded externally by the epidermis and internally by the vascular system.

Cotyledon: the leaves of a plant embryo, which in some plants enlarge and function as a storage site for nutrients to support early growth after seed germination.

Crossing over: the exchange of corresponding chromatid segments (genetic material) of homologous chromosomes during synapsis in the first phase of meiosis.

Cyanobacteria: photosynthetic prokaryotes that have chlorophyll and release oxygen; sometimes referred to as blue-green algae.

Cytoplasm: cellular contents exclusive of the nucleus.

D

Deciduous plant: a plant that seasonally sheds its leaves.

Denitrifying bacteria: single-cellular organisms of the Bacteria domain that convert nitrate to atmospheric nitrogen.

Detritus: non-living organic matter important in the nutrient cycle in soil formation. When abundant, organic detritus is often known as *humus*.

Diatoms: aquatic unicellular algae characterized by a cell wall composed of two silica-impregnated valves.

Dicot: a kind of angiosperm characterized by the presence of two cotyledons in the seed; also called *dicotyledon*.

Diffusion: movement of molecules from an area of greater concentration to an area of lesser concentration.

Dihybrid cross: a breeding experiment in which parental varieties differing in alleles for two traits are crossed.

Dimorphism: two distinct forms within a species, with regard to size, color, organ structure, etc.

Division: a major taxonomic grouping of plants that includes classes sharing certain features with close biological relationships. Most biologists use the term phylum rather than division.

Dominant: a hereditary characteristic that is expressed when the genotype is homozygous or heterozygous.

Dormancy: a period of suspended activity and growth.

Double helix: a double spiral used to describe the three-dimensional shape of DNA.

E

Ecology: the study of the relationship of organisms and the physical environment and their interactions.

Ecosystem: a biological community and its associated abiotic environment.

Embryo: a plant at an early stage of development. An embryo develops from a zygote and may begin growth immediately or become dormant.

Endosperm: a plant tissue of angiosperm seeds that stores nutrients; the endosperm of an angiosperm is typically $3n$ in chromosome number. It is produced by fusion of a sperm and polar nuclei.

Epicotyl: portion of a plant embryo that develops to become part of the stem, above a cotyledonary node.

Epidermis: the outermost protective layer of cells of a plant.

Epiphyte: nonparasitic plant that grows on the surface of other plants.

Estuary: the mixing zone between fresh water and seawater at the mouth of a river.

Evolution: genetic and phenotypic changes occurring in populations of organisms through time, generally resulting in increased adaptation for continued survival of a species. Evolution may also result in extinction.

F

Fertilization: the fusion of two haploid gametes to form a diploid zygote.

Fibrous root system: a mass of roots of about equal size.

Filament: a long chain of cells.

Filtration: the passage of a liquid through a filter or a membrane.

Flora: a general term for the plant life of a region or area.

Flower: the blossom of an angiosperm that contains the reproductive organs.

Fossil: any preserved ancient remains or impressions of an organism.

Frond: the leaf of a fern, cycad, or palm containing many leaflets.

Fruit: a mature ovary enclosing a seed or seeds.

Fruiting body: a reproductive structure of a fungus or slime mold in which spores are produced.

G

Gamete: a haploid sex cell, often a sperm or egg.

Gametophyte: the haploid, gamete-producing phase in a life cycle of a plant.

Gemma *(pl. gemmae)*: a small vegetative outgrowth of the thallus in liverworts or certain fungi that can develop into a new organism.

Gene: part of the DNA molecule located in a definite position on a certain chromosome and coding for specific protein product.

Gene pool: the total of all the alleles of the individuals in a population.

Genetic drift: evolution by chance process; often due to the loss of parts of a population.

Genetics: the study of genes, gene products and heredity.

Genotype: the genetic makeup of an organism.

Genus: the taxonomic category above species and below family; the first name of a scientific binomial.

Germ cells: gametes or the cells that give rise to gametes.

Germination: the process by which a spore or seed ends dormancy and resumes metabolism, development, and growth.

Gibberellin: plant hormone producing increased stem growth by promoting cell division; also promotes seed germination and flowering.

Girdling: removal of a strip of bark from around a tree down to the wood.

Grana: a "stack" of flattened membrane disks (thylakoids) within the chloroplast that contain chlorophyll.

Gravitropism: plant growth oriented with respect to gravity; stems grow upward, roots grow downward; also called *geotropism*.

Growth ring: an annual growth layer of secondary xylem (wood) in gymnosperms or angiosperms.

Guard cells: epidermal cells at the side of a stomate that help to control the stoma size.

Gymnosperm: a vascular plant producing naked (exposed) seeds, as in conifers.

Gynoecium: the carpel or carpels of an angiosperm flower.

H

Habitat: the ecological abode of a particular organism.

Herbaceous: a non-woody plant or plant part.

Herbaceous stem: stem of a non-woody plant; stem lacking wood.

Heredity: the transmission of certain characteristics, or traits, from parents to offspring via the genes.

Heterozygous: having two different alleles for a given trait.

Holdfast: basal extension of a multicellular alga that attaches it to a solid object.

Homothallic: species in which individuals produce both male and female reproductive structures and are self-fertile.

Hybrid: an offspring from the crossing of genetically different strains or species.

Hypha: a single filament of cells that makes up the vegetative body of a fungus.

Hypocotyl: portion of a plant embryo that contributes to the root development. The hypocotyl is below the cotyledonary node.

I

Indigenous: organisms that are native to a particular region; not introduced.

Internode: region between stem nodes.

K

Karyotype: the number and type of chromosomes characteristic of the species or of a specific individual.

Kingdom: a taxonomic category grouping related phyla.

L

Lateral bud: a bud often in a leaf axis that has the potential to become a branch or other structure.

Lateral root: a secondary root that arises by branching from an older root.

Leaf: a lateral appendage from a plant that has the principal function of photosynthesis.

Leaf veins: plant structures that contain the vascular tissues in a leaf.

Legume: a member of the pea, or bean, family; also the fruit of this family.

Lenticel: spongy area in the periderm of a stem or root that permits interchange of gases between internal tissues and the atmosphere.

Lichen: an alga and fungus forming a single thallus and coexisting in a symbiotic relationship.

Locus: the specific location or site of a gene within the chromosome.

M

Marine: pertaining to the sea or ocean.

Medulla: the center portion of an organ.

Megaspore: a plant spore that will germinate to become a female gametophyte.

Meiosis: nuclear division by which haploid nuclei are formed from a diploid nucleus; also referred to as reduction division.

Meristem: undifferentiated plant tissue that is capable of dividing and producing new cells.

Mesophyll: the ground tissue layer of a leaf containing cells that are active in photosynthesis, gas exchange and sometimes storage.

Microspore: a spore that develops to produce the male gametophyte.

Migration: movement of organisms from one geographical site to another.

Mitosis: the process of nuclear division, in which the two daughter nuclei are identical and contain the same number of chromosomes; often followed by cell division.

Monocot: a type of angiosperm in which the seed has only a single cotyledon; also called *monocotyledon*.

Mutation: a variation in heritable characteristic caused by a change in DNA; a permanent transmissible change in which the offspring differ from the parents.

Mutualism: a beneficial relationship between two organisms of different species.

Mycelium: the mass of hyphae that constitutes the body of a fungus.

N

Natural selection: the evolutionary mechanism by which organisms with adaptive traits pass on their genes to the next generation.

Nitrogen fixation: a process carried out by certain prokaryotes, such as some soil bacteria, whereby free atmospheric nitrogen is converted into ammonia compounds.

Node: location on a stem where a leaf is attached.

Nucleus: a spheroid body within the eukaryotic cell that contains the chromosomes of the cell.

Nut: a hardened and dry single-seeded fruit.

O

Oogonium: a unicellular female reproductive organ of some non-vascular plants and fungi that contains a single or several eggs.

Organ: a structure consisting of two or more tissues, which performs a specific function.

Organelle: a minute structure of the eukaryotic cell that performs a specific function.

Organism: an individual living creature.

Osmosis: the diffusion of water from a solution of lesser concentration to one of greater concentration through a semipermeable membrane.

Ovule: the female reproductive structure in a seed plant that contains the megasporangium where meiosis occurs and the female gametophyte is produced. Ovules mature to become seeds.

P

Paleobotany: the study of fossil plants.

Palisade layer: the columnar layer of the mesophyll of a leaf where abundant photosynthesis occurs.

Parasite: an organism that resides in or on another from which it derives sustenance.

Parallel evolution: the development of similar adaptive traits in different species as a result of similar selective pressures.

Parasite: an organism that derives nutrients from another species; an endoparasite lives within the host organism and an ectoparasite lives on the host organism–both relationships are detrimental to the host organism.

Parenchyma: the principal structural cells of herbaceous plants; a relatively non-differentiated plant cell type characterized by thin primary cell walls.

Passive transport: molecular movement across a membrane that does not require energy.

Pathogen: a disease causing organism.

Pectin: an organic compound in the intercellular layer and primary wall of plant cell walls; the basis of fruit jellies.

Pedicel: the stalk of a flower in an inflorescence.

Perennial plant: a plant that lives throughout the year and grows during several to many growing seasons.

Perfect flower: a flower having stamens and carpels contained in the same organ.

Pericarp: the fruit wall that forms from the wall of a mature ovary; or female gametophyte tissue enclosing tetrasporophyte in some red algae.

Pericycle: a tissue in the roots and in the stems of certain plants that is bounded externally by the endodermis and internally by the xylem and phloem.

Petal: modified leaf occurring in a flower. Petals are often colored and functional in attracting pollinators: collectively called the *corolla*.

Petiole: structure of a leaf that connects the blade to the stem.

Phenotype: the appearance of an organism created by the genotype and environmental influences.

Phloem: vascular tissue in plants that transports nutrients.

Photosynthesis: the process of using the energy of the sun to make carbohydrates from carbon dioxide and water.

Phototropism: plant growth or movement in response to a directional light source.

Phycology: the study of algae.

Phylogeny: the evolutionary relationship among organisms.

Phytoplankton: microscopic, free-floating, photosynthetic organisms that are the major primary producers in fresh-water and marine ecosystems.

Pistil: a reproductive structure of a flower comprised of the stigma, style, ovary; and one or more carpel.

Pith: a centrally located tissue within a dicot stem.

Plankton: aquatic free-floating microscopic organisms.

Plastid: an organelle of a plant where photosynthesis or food storage occurs.

Pollen grain: a mature microspore containing the male gametophyte generation of seed plants.

Pollen tube: a passageway formed after germination of a pollen grain that permits passage of male gametes into an ovule.

Pollination: the delivery by wind, water, or animals of pollen to the stigma of a seed plant leading to fertilization.

Population: all the organisms of the same species in a particular location.

Producers: organisms within an ecosystem that synthesize organic compounds from inorganic constituents.

Prokaryote: organism, such as a bacterium, that lacks the specialized organelles and a nuclear envelope characteristic of complex cells.

Prothallus: the gametophyte generation of a nonvascular plant.

Protonema: the first stage of gametophyte development in mosses and liverworts.

R

Radial symmetry: symmetry around a central axis so that any half of an organism is identical to the other.

Receptacle: the tip of the axis of a flower stalk that bears the floral organs.

Regeneration: the regrowth of tissue or the formation of a complete organism from a portion.

Renewable resource: a commodity that is not used up because it is continually produced in the environment.

Replication: the process of producing a duplicate; DNA is replicated prior to cell division.

Rhizoid: a minute hairlike extension of a fungus or plant that functions in nutrient and water absorption.

Rhizome: an underground stem in some plants that stores photosynthetic products and gives rise to above-ground stems and leaves.

Root: the anchoring subterranean portion of a plant that permits absorption and conduction of water, minerals, and nutrients.

Root cap: end mass of parenchyma cells that protects the apical meristem of a root.

Root hair: epidermal projection from the root of a plant that functions in absorption of water the nutrients.

S

Salinity: saltiness in water or soil; a measure of the concentration of dissolved salts.

Sap: the fluid content of the xylem or the sieve elements of the phloem.

Saprophyte: a heterotrophic bacterium, fungus, or plant that absorbs nutrients directly from dead and decaying organic matter.

Savanna: open grassland with scattered trees.

Sclerenchyma: supporting tissue in plants composed of cells with thickened secondary walls.

Secondary growth: plant growth in girth from secondary or lateral meristems.

Seed: a plant reproductive body developed from a matured ovule and consists of a plant embryo with a food reserve enclosed in a protective seed coat.

Seed coat: the outer protective epidermal layer of a seed.

Seedling: a developing sporophyte, which develops from a germinating seed.

Sepal: outermost whorl of flower structures beneath the petals; collectively called the *calyx*.

Sessile: organisms that lack locomotion and remain stationary.

Sexual reproduction: the fusion of a male and female gamete, followed by meiosis and genetic recombination at some portion of the developmental life cycle.

Shoot: portion of a vascular plant that includes a stem with its branches and leaves.

Shrub: a relatively short bushy woody plant that generally has several stems arising from or near the ground.

Sieve tube: a linear group of cells in the phloem functioning in translocation of dissolved photosynthetic products.

Simple fruit: a ripened ovary derived from one carpel or several united carpels.

Simple leaf: a continuous, undivided laterally extending organ of a plant; opposed to a compound leaf.

Somatic cells: all the cells of the body of an organism except the germ cells (gametes).

Sorus: a cluster of sporangia on the underside of fern pinnae (leaflets).

Species: a group of morphologically similar organisms that share a gene pool and are capable of interbreeding and producing fertile offspring and are generally reproductively isolated from other species.

Sperm: a mature haploid (1*n*) male gamete.

Spermatophyte: a seed plant.

Spirillum *(pl. spirilla)*: a spiral-shaped bacterium.

Spongy parenchyma: leaf tissue containing loosely arranged, chloroplast-bearing cells.

Sporangium: any structure within which spores are produced.

Spore: a reproductive cell capable of developing into an adult organism without fusion with another cell.

Sporophyll: a sporangium-bearing leaf.

Stamen: a reproductive structure of a flower, comprised of a filament and an anther, where pollen grains are produced.

Starch: carbohydrate molecule synthesized from photosynthetic products; common food storage substance in many plants.

Stele: the primary vascular tissue at the central core of a root or stem.

Stem: the supporting axis of a vascular plant either above ground, or in some plants (such as those having rhizomes or corms), below the ground as well.

Stigma: the upper portion of the pistil of a flower. Pollen grains become attached to the stigma.

Style: the long slender portion of the pistil of a flower.

Succession: the sequence of ecological stages by which a particular biotic community gradually changes until replaced by another community.

Succulent: a fleshy plant with fluid-storing stems and leaves.

Sucrose: a disaccharide (double sugar) consisting of a linked glucose and fructose molecule: the principal transport sugar in plants.

Superior ovary: a type of flower where the ovary is free and separate from the calyx.

Symbiosis: a close association between two organisms where one or both species derive benefit.

Syngamy: union of gametes in sexual reproduction; *fertilization*.

T

Taproot: a plant root system in which a single root dominates the root system.

Taxon: a taxonomic grouping, such as species, genus, class, order, or phylum.

Taxonomy: the science of describing, classifying, and naming organisms.

Thallus: a flattened plant body often with little tissue specialization and lacking roots, stems, or leaves.

Tuber: a thickened underground stem, such as a potato.

Tendril: a modified fleshy portion of a plant into a slender coiled structure that aids in the support of the stem; found only in certain angiosperms.

Tissue: an aggregation of similar cells and their binding intercellular substance joined to perform a specific function.

Toxin: a poisonous compound.

Trait: a distinguishing feature of an organism, often studied in heredity.

Turgor pressure: osmotic pressure that provides rigidity to a cell.

U

Unicellular: an organism consisting of a single cell.

Unisexual: in terms of botany, a flower lacking either stamens or carpels; a perianth may be present or absent.

V

Vacuole: a membrane-bound, fluid-filled organelle.

Variety: a division of plants or animals below the subspecies level.

Vascular cambium: a layer of meristematic tissue in roots and stems of many vascular plants that continues to produce secondary vascular tissue.

Vascular tissue: plant tissue composed of xylem and phloem, functioning in transport of water, nutrients, and photosynthetic products throughout the plant.

Vascular plant: a plant that has the vascular tissue xylem and phloem.

Vegetative: plant parts not specialized for reproduction; asexual reproduction.

Viable: the ability to survive.

W

Weed: a popularized term referring to a herbaceous plant lacking in commercial or aesthetic value and living with and hindering the growth of desirable plants.

Wood: the interior tissue of a tree composed of secondary xylem.

X

Xerophyte: a plant adapted to live in an arid environment.

Xylem: vascular tissue in plants that transports water and minerals.

Z

Zoospore: a flagellated motile plant spore.

Zygote: the union of haploid gametes ($1n$) in the formation of a diploid ($2n$) cell; a fertilized egg.

Index

SUFFOLK UNIVERSITY

MILDRED F. SAWYER LIBRAR

8 ASHBURTON PLACE

BOSTON, MA 02108